I0055749

COAL
THE AUSTRALIAN STORY

Connor Court Publishing Pty Ltd
PO Box 7257
Redland Bay QLD 4165

sales@connorcourt.com
www.connorcourtpublishing.com.au
Phone 0497 900 685

ISBN: 9781925826609

Front Cover Design: Maria Giordano
Front Cover Photo: courtesy of National Archives of Australia. Miner with coal cutting
machine in Burwood Colliery Newcastle 1946. A1200, L6941
Printed in Australia

COAL

THE AUSTRALIAN STORY

From convict mining to the birth of a world leader

Denis Porter

Connor Court Publishing

List of shortened forms

AACo	Australian Agricultural Company
ABARE	Australian Bureau of Agricultural and Resource Economics
ABS	Australian Bureau of Statistics
ACA	Australian Coal Association
ACIRRT	Australian Centre for Industrial Relations Research and Training
ACSA	Australian Collieries Staff Association (now part of Professionals Australia)
AGPS	Australian Government Publishing Service
AIRC	Australian Industrial Relations Commission
ALP	Australian Labor Party
AMIC	Australian Mining Industry Council; now the Minerals Council of Australia
AMWU	Amalgamated Metal Workers Union
ANU	Australian National University
APESMA	Association of Professional Engineers, Scientists and Managers Australia; now Professionals Australia
AusIMM	Australasian Institute of Mining and Metallurgy
AWU	Australian Workers' Union
BCA	Business Council of Australia
BHP	Broken Hill Proprietary Company Limited; now BHP Group Limited
BHPB	BHP Billiton Limited
BIE	Bureau of Industry Economics

BMA	BHP Mitsubishi Alliance
BREE	Bureau of Resource Economics and Energy
CFMMEU	Construction Forestry Maritime Mining and Energy Union
CIT	Coal Industry Tribunal
CQCA	Central Queensland Coal Associates
CRA	Conzinc Rio Tinto of Australia Limited
DBCT	Dalrymple Bay Coal Terminal
EIA	Energy Information Administration; part of US Department of Energy
EIS	Environmental impact statement
EPDC	Electric Power Development Company of Japan
ETU	Electrical Trades Union
Federation	Miners' Federation; now the Mining and Energy Division of the CFMMEU
FEDFA	Federated Engine Drivers' and Firemen's Association
IEA	International Energy Agency
JABAS	J&A Brown and Abermain Seaham Collieries Limited
JCB	Joint Coal Board
JFY	Japanese financial year; commences 1 April
JPU	Japanese power utilities
JSM	Japanese steel mills
LCA	Local Coal Authority
MIM	Mount Isa Mines Limited
MITI	Japanese Ministry of International Trade and Industry; now Ministry of Economy, Trade and Industry
NCIG	Newcastle Coal Infrastructure Group
NSW	State of New South Wales, Australia
NSWCCPA	NSW Combined Colliery Proprietors' Association

NSWMC	NSW Minerals Council
OECD	Organisation for Economic Cooperation and Development
PKCT	Port Kembla Coal Terminal
PWCS	Port Waratah Coal Services
QCA	Queensland Coal Association
QCB	Queensland Coal Board
QMC	Queensland Minerals Council; now Queensland Resources Council
QRC	Queensland Resources Council
Qld	State of Queensland, Australia
SA	State of South Australia, Australia
TDM	Thiess Dampier Mitsui Coal Pty Limited
TPM	Thiess Peabody Mitsui Coal Pty Limited
WA	State of Western Australia, Australia

Foreword

The Australian coal industry has been the subject of many books, papers and university theses, but there has been no recent book to bring together the overall story of the development of the industry in the two major coal exporting states – NSW and Queensland. Denis Porter's book now does this in a way that the general reader as well as those acquainted with the industry will find informative and highly readable.

The coal industry, as those who have worked in it will know, has been characterised by its ups and downs and industrial turmoil, but has emerged from its turbulent past as a strong and competitive industry, competing on the world stage with other major producing countries. While it is now the leading exporter of metallurgical coal and the second largest exporter of thermal coal, the industry had humble beginnings in Newcastle with convicts mining the coal under horrific conditions. Denis traces the development of the industry through the 1800s to the time of Federation and the early 1900s, through the dark years of the 1920s and 1930s and the Second World War and the following two decades. The industry emerged from the Second World War in a poor condition and with negligible exports, but by the 1960s was facing the challenge of becoming a major supplier to the Japanese steel mills, competing with the US producers who were the dominant suppliers. The competition between the underground producers in NSW and the emerging open cut producers in the Queensland Bowen Basin was also beginning and would become more intense during the 1970s. By the end of the 1960s, the Australian industry's future as the leading supplier to the Japanese was by no means assured, but as the title of this book indicates, that decade can be seen as the time which marked the birth – or perhaps the re-birth – of the coal industry we know today.

Coal: the Australian Story – from convict mining to the birth of a world leader is a story which deserves to be better understood by all Australians. It is a story, not only of turmoil, but also of perseverance, major reforms and restructuring, and a story which involves an industry which has been of fundamental importance to the Australian economy for most of the industry's life.

The coal industry today faces major challenges, with thermal coal in particular under the spotlight and with many questioning its long term role in power generation. Those challenges will be resolved over the coming years, but they should not detract from Australians having a better understanding of the industry's rich history.

I commend Denis for his own very positive involvement in the NSW coal industry, for his diligence, memory and undoubted skill in researching and recording a most important and interesting Australian industry. A book I heartily recommend.

A J (Tony) Haraldson AM

Various roles in R W Miller associated companies, 1963-1988.

Chief Executive, Coal & Allied Industries Ltd, 1989 to 1993

Chief Executive, Coal Operations Australia Limited/Billiton Coal Australia, 1994 to 2001

Alternating Chairman/Director, Coal Services Pty Limited 2004 to 2014

Chairman, Hunter Valley Coal Chain Coordinator 2009-2017

Table of Contents

Introduction

Coal's role in post 1788 Australia dates back to the very early days of colonisation by the British – and there is even a close link to 1770 when Cook arrived. Captain James Cook (or Lieutenant Cook as he then was) had been intimately involved in the British coal industry as a young man. He was born in Yorkshire in the heart of coal country and served his apprenticeship on colliers - ships used for the coal trade - on the British North Sea coast. And the *Endeavour*, the ship on which Cook made his voyage to Australia, began its life as a collier. Coal does not figure prominently in Australian history prior to the arrival of the British in 1788. The Awabakal people, the Aboriginal people who inhabited the Lake Macquarie and Newcastle region in New South Wales, are thought to be the only people to discuss coal in their legends and are thought to have burnt coal in their fires. According to the Reverend Lancelot Threlkeld, who established a mission for the Aboriginal people in present day Belmont, the name for the Lake Macquarie district was Nikkin-bah, or place of coal.[1]

The first coal produced after the colony was established by the British in Australia was mined in Newcastle in 1798. The first coal exported from Newcastle may have been in 1799 on board *The Hunter* and bound for Bengal via Sydney; or it may have been two years later, but in any case, exports featured prominently in those early years. The Australian coal industry is a now a world leader. Australia is the world's largest coal metallurgical coal exporter, and the number two thermal coal exporter, and in recent years coal has been Australia's second largest export commodity earner behind iron ore. It was the leading export earner in calendar year 2018. The industry is technologically advanced, employing many highly skilled workers, is highly productive, environmentally responsible and has a good safety record, particularly compared with coal industries in other countries. However if we go back to World War

Two and the early post war years, it was a dramatically different story
– the industry was wracked by industrial strife, incapable of producing
sufficient coal for the country's needs, exporting only tiny percentage
of its production, largely unprofitable, and with a safety record which,
if recorded today, would see many senior executives and directors
threatened with incarceration for failing in their duty of care to their
employees.

This book looks at the industry from the early days of the Colony,
through the period to the Second World War, then to the post war period
when the industry began to be modernised, and finally to the 1960s
when the industry was starting out on its transformation to one which
would be increasingly focussed on the export market. The period from
the 1970s to today will be the subject of a second volume of *Coal: the
Australian story*.

From its beginnings in Newcastle, the industry grew through the
1800s to become a critical part of Australian life, with coal powering
our factories, railway locomotives and steamships, and fuelling the plants
which produced gas for street lighting, lighting in homes and for use by
various industries. During the 1900s, coal was the energy source for the
many new power stations built throughout the nation to generate the
electricity which became such an essential part of our everyday life. By the
time of Federation the coal industry, although modest in size compared
with the huge industries in countries such as the UK, USA and Germany,
had also developed important export markets, although by World War
Two, the level of exports had fallen away to almost nothing. Coal also
was the basis for the development of our iron and steel industry, with
BHP's first steelworks established in Newcastle in 1915, followed in 1928
by the AI&S steelworks in Port Kembla.

While coal mining has been an important part of the development
of Victoria, South Australia, Western Australia and Tasmania, this book
focusses on NSW and Queensland as these two states have by far the
largest coalmining industries and have become major coal exporting
states. It has also been in these two states where the major industrial
battles have been fought, where foreign investment has played such a

major role in the industry's development and often subject of much controversy. The coal industries in NSW and Queensland have also seen unique or special regulation through coal boards and the industry's industrial tribunal which was separate from the mainstream industrial relations system for so many years.

Victoria's brown coal mines, developed from the 1920s, continue to fuel much of that state's electricity; its black coal mining ceased in 1968 with the closure of the Wonthaggi mine. Coal mining ceased in Leigh Creek in South Australia in 2015 when the last coal-fired power stations in that state closed. Mining of black coal continues in Collie Western Australia, with the coal used for power generation, and in Tasmania, where coal is mined in the Fingal Valley for use in cement production. Those states have their own rich stories, some of which remain to be told.

For much of the past two centuries, the coal industries in NSW and Queensland developed as quite distinct and separate industries, although in the 1800s there was some interstate trade in coal, mostly involving sales from NSW into Queensland. The Queensland industry, beginning in the Ipswich area and later expanding into other areas of the state, was essentially operated to supply domestic needs. Exports only began to be important in the 1960s with the pioneering work in the Bowen Basin by Thiess Brothers in conjunction with Mitsui and by Utah. NSW was involved in exporting coal from its early days, but particularly from the late 1800s, and by 1900 its exports had grown to around 1.3 million tonnes a year, with sales to what were termed "intercolonial ports" (other Australian colonies/ states and New Zealand) a further 1.8 million tonnes. Those exports to other states were providing vital coal for example to Melbourne for its gas industry and railways, and to the South Australian railways and the Perth Gas Company. Exports to Victoria were to drastically reduce however following a major industrial dispute in NSW in 1909 which cut supplies and led to the Victorian Government deciding to develop its own coal mining industry in Wonthaggi.

The NSW and Queensland industries largely continued along their separate paths until World War Two, when the Commonwealth

Government assumed control of the industry through wartime powers. Queensland established its own coal boards in the 1930s, and but for the change of government in NSW in 1932, a coal board in that state would have also been in operation that year or shortly after. After the end of the war, with the industry in need of drastic change, the Joint Coal Board was established in NSW with sweeping powers to control and regulate. Queensland however wanted no part of the unique new joint Commonwealth-NSW arrangements and set up its own Queensland Coal Board. Both Boards would continue to operate until the 1990s.

The post war years saw the industries in NSW and Queensland take major steps to modernise, although some major companies in NSW had already gone well down the path to mechanise during the 1930s. The industry in NSW was much larger than in Queensland and included big companies such as BHP and its AI&S subsidiary, JABAS, Caledonian and the Joint Coal Board itself which rivalled BHP/AI&S in size as a producer by the early 1950s. Queensland was dominated by small companies, many of which were family operations, with many also unsuited to mechanisation. Many of those small operations would close over the coming years, unable to be part of the developing new industry.

While the industry in both states expanded in the post war years, supplying much needed coal for a rapidly growing population and economy, there would also be major structural changes severely impacting on jobs and local communities. The decline of the South Maitland coalfield in the late 1950s saw the Miners' Federation call for action to fix what it said was a crisis in the industry. The mines producing coal from the Greta seams on this coalfield, in the area between Maitland and Cessnock, were confronted by major changes in the market for their coal, changes which were irreversible. Greta seam coal was sold to the railways, but the railways were switching from steam powered to diesel powered locomotives. Greta coal was also sold to AGL for gas production, but AGL was switching to buying gas from a nearby petroleum refinery, and by 1976 Sydney would be supplied with natural gas by the pipeline from Moomba in South Australia.

It was not until the 1960s that the parallel development of the industry in NSW and Queensland began to cause problems for the regulators and for the producers, particularly those in NSW. The new and large scale developments in the Bowen Basin would be in direct competition with the established producers in NSW and would be lower cost operations. High cost operations in NSW would come to fear their new competitors north of the border, and the new customers, the Japanese steel mills, would enjoy the benefits of the new open cut mines in the Bowen Basin. Much of that story would then play out in the 1970s and beyond and will be part of the story in the second volume.

The 1950s and 1960s also saw Japan start to look to Australia as a source of coking coal for its steel mills. The Japanese would be an integral part of the development of the industry, initially from the involvement of Mitsui in exploration for coal and the development of the Moura and Kianga mines and the involvement of Mitsubishi as the conduit for Utah and its contracts with the Japanese steel mills and its developments of its major Bowen Basin mines. And Japan's key role in the expansion of the Australian coal industry (and expansions in Canada and South Africa) would also see the parallel decline of its own large coal industry, one which during the war had a workforce of around 400,000, with several thousand of those workers prisoners of war from Australia and other countries.

The book does not pretend to be in any way a complete history of the industry in NSW and Queensland; rather it is attempt to bring together some of the major events in the industry and some of the major influences on the way it has evolved and grown. There are no doubt events and stories that some readers will say should have been included and they may well be right. However I hope that the book can be seen as a credible overview of the industry and its history, and that it is does justice to an industry which, for all its problems and turmoil, has been critically important for the development of Australia. The book ends with the decade of the 1960s, a period when the industry was beginning in earnest its transformation into a modern industry, largely focussed on export markets. The second volume will take up the story from the turbulent 1970s, a decade of the oil shocks, inflation and currency realignments. That decade was also one which would see the Japanese steel industry's production reach its peak after growing rapidly over the previous two decades, but also would herald the expansion of the Australian thermal coal sector to supply new coal-fired power stations built to reduce many countries' dependence on oil. The 1980s and 1990s would also be turbulent decades for the industry, seeing major changes and controversies over coal prices and coal marketing, the industry's regulatory system and mine safety. The new century has then seen the historic impact of the growth of China on world commodity markets and industries, the debates over climate change and the future role of coal. The history of the coal industry in Australia has never been bland and boring.

1

Colonisation and the first century

Mary and William Bryant were convicts from Cornwall in England; both were sentenced to transportation for seven years, arriving on the *Charlotte* as part of the First Fleet in 1788. Mary gave birth to a child on board the ship and she and William married in February shortly after arriving in the new Colony. William was an experienced fisherman and was able to steal a small boat and, with Mary, two children and seven other convicts, headed north out of Port Jackson on a dark night on 28 March 1791. After an epic voyage of over 5,000 kilometres lasting 69 days they reached the island of Timor where they were arrested and imprisoned. One of the group, James Martin, recorded the journey, and referred to their departure from Sydney in an open boat, reaching a little creek after two days' sailing, and their discovery of a "quantity of fine burng (burning) coal."[2] The creek is believed to have been in the area between the entrance to Lake Macquarie and the Hunter River. Whoever did the cooking that night, may well have been the first European in Australia to cook with coal.

So the first discovery of coal in Australia by the British is believed to have been by this group of escaping convicts on 30 March 1791. The group's discovery was not known in the Colony for years to come. In the meantime, two "official" discoveries were made in 1797, coal was also found north of Sydney in 1796, and there was another likely discovery in 1796. Coal was also discovered in Tasmania in 1793 by the French explorer Labillardiere. In June 1796 David Collins, the Judge-Advocate, recorded that a fishing boat had brought back to Sydney some coal found near the beach on its return from Port Stephens.[3] George Bass and Matthew Flinders landed north of Wollongong on 28 March 1796 in the vessel Tom Thumb and found what we can assume to be coal: "The shore was mostly high with cliffs, and under the cliffs were lying black lumps, apparently of slaty stone, rounded by attrition."[4]

In 1797 a group returning to Sydney, after having been shipwrecked in Victoria, found coal on the beach near Coalcliff south of Sydney, and used the coal to keep themselves warm overnight. Governor Hunter recorded this discovery in a letter to London.[5] He sent George Bass and two of the party who had found the coal to check it out. Bass found a seam of coal six feet deep in the face of a steep cliff and which extended for eight miles. He also said that this was not the only coal he and his party discovered in that area. Bass also predicted that the Blue Mountains extended to the coast in this area and that the coal seam he had seen ran through the mountain range.[6]

In another letter to London in 1798 Hunter referred again to the discovery and to samples of coal from the area which he sent to Sir Joseph Banks, and said that the coal was very good, but "difficult to attain, being a strata or vein of an immense steep cliff, near the sea…" He cautioned that "unless we can find some little harbour near, can we hope to derive any great advantage from it."[7] With no natural harbour nearby, the coal resources to the south of Sydney would therefore have to wait many years to be exploited.

Newcastle is discovered

The first discovery of coal which actually led to mining of coal was near the entrance to the Hunter River. In 1797 Lieutenant John Shortland was sent north from Sydney to look for some convicts who had escaped. He and his crew went as far as Port Stephens, but found no convicts. However on their way back they sheltered from a storm between Nobby's (then an island) and the mainland. Shortland then discovered the Hunter River and found coal near the entrance. He brought samples back to Sydney. The Hunter River and the local settlement established shortly after were known in the next few years as Coal River. Coal began to be extracted from that location in 1798. Mining was probably too fine a term to describe what initially was essentially hacking coal from the face of the hill.

The first export of coal may have been in 1799. A letter from Surgeon John Thomson refers to a ship named Hunter recently having returned to Bengal loaded with coal, and noted that "it gave no small satisfaction to every person interested in the prosperity of the colony to see the first export of it..."[8] However J W Turner's book on the history of coal in the Newcastle area notes that there is some doubt that this was in fact the first export.[9] We also have Governor King's reference to coal exported in the Earl Cornwallis in 1801 in a letter he wrote to Banks: "The first cargo of coals bought from the Coal River in a Government vessel I exchanged with the master of the Cornwallis, who goes to Bengal from hence for iron … I believe this is the first return ever made from New South Wales."[10]

John Platt - our first miner

King was keen for coal to be found in the area near the Sydney Cove colony and he employed the only miner he could find in the Colony to search for coal. Writing to London in 1801 he said that: "Ever since I took the command, an experienced miner with eleven men have been employed boring in the most likely places to produce coals in the neighbourhood…he has got (down) ninety-six feet, but no coals, only very thin veins..."[11] One of King's letters that year identified the area being explored as the head of the George's River. "He has opened a shaft 24 feet deep and has bored 50 feet …has passed two very thin strata of very fine coal, and …he is very confident of succeeding. If he should fail … I shall remove him and his men to the northward of the rivers..."[12]

The miner was John Platt, a convict from England who had been sentenced to life imprisonment in 1798 and arrived in the Colony in 1800. But Platt's optimism did not lead to any discovery and the George's River area did not become a mining area. Platt was also a member of the party led by Lt Colonel Paterson which was sent to the Hunter River to survey its resources in 1801. Platt found good quantities of coal which was shipped back to Sydney. Paterson returned to Sydney, but left Platt and other convicts behind to work on the coal seam. However the first official attempt to mine coal on the Hunter River with convict labour proved unsuccessful and the convicts were sent back to Sydney at the end of 1801. Mining continued spasmodically after the convicts were sent back to Sydney, with ships' crews doubling as miners.

Following a rebellion by Irish convicts at Castle Hill in March 1804, Governor King decided to establish a penal settlement on the Hunter River. Official mining was re-started in 1804, again using convict labour, and from this time Newcastle became a permanent settlement. The first shaft was sunk near what was later the site for the hospital in Watt Street. Newcastle was officially named on 24 March 1804. King also made private mining of coal illegal and ordered that experienced miners were to work in the mines with the other convicts.[13] This set the basis for the period to 1831, when the mining of coal in Newcastle would be exclusively government-run.

The mining that had been carried on before 1804 was obviously not of a high standard. Lieutenant Menzies, the officer who re-started the settlement with the party of rebellious Irish convicts and their guards noted that "the mines have hitherto been dug in a shameful manner. Never have (the coal cutters) been at the trouble of hewing supports, leaving them to fall in anyway." [14] Menzies found that the coal seam most suited for working was on Colliers Point, now known as Fort Scratchley. Production expanded steadily in the following two decades. In 1805 only 150 tons of coal were produced, but by 1820 this had grown to almost 4000 tons. Conditions for the convict workers in those early years were miserable, as evident from the following quotes.

"The mine workers lived in dormitories in double rows of cribs. Each of those measured 4 feet 3 inches by 6 feet (about 1.3 metres by 1.8 metres), and was occupied by three men. Some of the cribs were in rooms, some in sheds holding up to twenty-four men. The convicts rose at five and worked till eight, again from nine till noon and from two till sunset, save on Saturdays when they had the hours between 10 am and 4 pm to themselves. They were checked over four times a day lest they escape. Constables and overseers from their own number – rude, untrustworthy fellows who occasionally escaped themselves – supervised them."[15]

"After being identified by the Overseer, the men were singularly lowered into the mine by bucket. Their work was to fill the wagons with coal, drag them to the opening at the foot of the shaft and tip out the contents for haulage to the surface. They slept naked in part of the workings where the heat was excessive. No bedding was provided and the only sleeping comfort was afforded by scraping coal dust into heaps as a crude mattress. The convicts were taken to the surface on Saturday afternoons to wash themselves and their clothing in seawater, then marched to their barracks and confined there until they returned to the mine on the Monday morning."[16]

The few skilled miners were the ones who actually dug the coal. They were expected to extract two and a half tons a day and received double the food rations issued to other convicts in Newcastle. "For each skilled miner, there were several who bailed out water, wheeled the coal to the shaft in barrows, raised it by windlass and moved it to the wharf. For these workers … there were only the normal rations, viz, eight pounds of wheat and four pounds of salt pork or seven pounds of salt beef per man per week. Fresh meat was rarely issued and the lack of vegetables made scurvy a fairly common complaint in Newcastle." [17]

John Bigge recommends mines be privatised

In 1819 the British Colonial Secretary, Lord Bathurst, appointed John Bigge, a British judge, to hold an inquiry into the Colony. With the high crime rate in Britain still a major concern for the British Government, Bathurst's main objective was to have an inquiry into the extent to which the transportation of convicts to the colony acted as a deterrent; he saw Governor Lachlan Macquarie's policies towards convicts as much too compassionate and humanitarian. No doubt coal mining was not top of the list of issues on Bigge's mind when he landed in Sydney in September

1819 and when he quickly began to gather evidence before returning to Britain in early 1821.

But Bigge had a very wide-ranging brief, and as part of his inquiry he visited Newcastle and collected evidence from leaders in the settlement. Coal mining of course was the dominant activity in Newcastle at that time and Bigge's findings would become critical for the future course of the industry. Evidence in Newcastle to Bigge's inquiry gives some indication of the health of the convict miners and the work they carried out. Surgeon Evans's evidence indicates that the settlement had a higher illness and mortality rate than a more normal colonial settlement and that the cause was the poor diet, with coal mining also causing lung disease. "The foul air that is breathed there produces spitting of blood and difficulty of breathing."[18]

Benjamin Grainger was the coal mining superintendent in Newcastle at that time. In his evidence he said that there were eight hewers working in the mine. The mine shaft was approximately 34 metres deep, and the tunnel leading to the coal face was about 1.4 metres high. Asked whether the health of the miners suffered as a result of the conditions, he agreed: "I think they do on account of the wet and remaining in their wet clothes after they come up from the pit. They also suffer from want of a change of clothes."[19] Asked whether the convicts worked naked in the coal mine, Grainger said that they did, "except for a loose trouser."

Bigge's third report released in 1823, *The State of Agriculture and Trade in the Colony of New South Wales*, included his observations on the Newcastle settlement and concluded that: "The labour in the coal mines at Newcastle is found to be prejudicial to the health of the convicts, on account of the bad air that they breathe, and the difficulty that is experienced in clearing it of water. Asthmas, pulmonary and rheumatic complaints are those from which the miners most suffer." [20] Bigge's report recommended that the NSW Government mines in Newcastle be leased to private operators. A process then ensued in the 1820s which culminated in the Government and the Australian Agricultural Company (AACo) agreeing that the company would take over operation of the mines.

The Australian Agricultural Company monopoly

The AACo was a company established in 1824 under an Act of the British Parliament to invest in wool and other pastoral ventures: "An Act …for granting certain powers and authorities to a Company to be incorporated by Charter to be called 'The Australian Agricultural Company' for the cultivation and improvement of waste lands in the Colony of New South Wales and for other purposes relating thereto." AACo's directors represented some of the cream of British society and included the Attorney General, Members of Parliament, the Governor of the Bank of England, several directors of the British East India Company, as well as leading bankers and other business people. Coal initially was not part of its plans, but the AACo was to become a major part of the coal industry for the rest of the 1800s and into the early years of the 1900s.

In 1826 the directors of AACo reported that the British Government had agreed to lease the Newcastle coal mines for 31 years under certain conditions. The chairman said that the mines had been operated "on a very limited scale and in a very imperfect manner by Government… until the year 1817, by a simple adit, or opening of the surface of the cliff, which served as a drift, both to carry off the water, and to wheel away the coals."[21] Since 1817 the operation, he said, had been through a perpendicular shaft around 36 metres deep, with convicts operating a windlass, and with a second shaft recently sunk, giving access to deeper seams and producing higher quality coal. The AACo directors were optimistic about the market for the coal, with Sydney already a significant market and smaller towns in the colony looking promising; overseas markets including in India, Batavia (Indonesia), Canton (China) and Singapore which had been supplied by Britain were also seen as potential customers. The likely expansion of steamships burning coal into the region was also seen as a positive. However it took another five years before AACo actually took over and opened its new mine in Newcastle.

The British Colonial Secretary sent instructions to Governor Darling in 1828 detailing the elements of the agreement with AACo for the takeover of the mines and the allocation of new leases. AACo was granted two large areas of land in Newcastle on which it could mine, one of 1500 acres and the other 500 acres. The British Government granted AACo a virtual monopoly; in the words of the Colonial Secretary: "His Majesty's Government deem it proper to desire that, for the next 31 years, no Governor will grant or convey any Coal Mines, or land containing any coal mine, without a specific exception of coal in such grant or conveyance, not afford any assistance in convict labour for the working of any coal mine, to any other company, or to any individual or individuals without the previous sanction of the Government at home; a sanction which would probably be granted, if the company should avail themselves of their monopoly to impose an exorbitant price upon coal, the produce of their mines."[22] However AACo's local officials were reluctant to proceed as they were not confident about the new venture and it was not until Sir Edward Parry arrived in December 1829 to assume control of the company's local operations that things began to progress.[23] One of Parry's major concerns was to firm up on the area to be chosen for the 500 acre lease and in August 1830 he advised the Chief Secretary that land to the west of the Government mines looked promising as the position for the lease. Parry also told the Chief Secretary that the Government's coal operations were useless to AACo in view of the poor quality of the coal; these operations were subsequently excluded from the 500 acre lease.[24] AACo began its own coal mining operations in 1831 and the official opening of the new mine and wharf took place on 10 December of that year. The new AACo mine had a main shaft which led to the coal seam and a second smaller shaft for ventilation; tunnels from the smaller shaft went in two directions, one towards the main shaft and the other to the side of the hill next to the "sea" (the harbour) through which the water from the main shaft was to be discharged. There was a railway leading from the mouth of the pit to the wharf which had facilities for loading the coal into ships and where the harbour was deep enough for the largest vessels expected to dock.[25]

The agreement with AACo recognised that the company would use a large number of convicts for its mining operations, and this was to be the case until the 1840s. The company was assigned over 200 new convicts from 1835 to 1841, but with the cessation of convict transportation to the colony in 1840, and the unwillingness of workers in the colony to accept underground work, AACo increasingly looked to its own immigration program. In 1839 AACo employed a total of 104 men, 95 of whom were convicts, and only 9 free men. This total grew to 183 in 1841, made up of 120 convicts and 63 free men. The company's numbers fell away in the next few years, but by 1847 there were only 4 convicts employed out of a total of 104, the rest being 52 free men and 48 who had been given a "ticket of leave" (ie were allowed freedom to live and work in the colony before the completion of their sentence).[26]

British miners recruited

In 1840 AACo recruited a group of miners from Wales, Cornwall, other English counties and Scotland. The majority went on strike shortly after they arrived. AACo's Commissioner, Captain King, was not impressed with the recruits, saying that he had never met "a more impertinent set of rogues" and that most of them "have brought with them a spirit of insubordination that will be difficult to subdue." But King was impressed with the Scots, whom he considered efficient miners, in contrast to the "Welsh" (King seemed to classify most of the non-Scottish miners as Welsh) who he said were "the very refuse of bad characters, discarded from various collieries, idle, drunken and discontented, suspicious and litigious."[27] The company then concentrated on recruiting Scots, and used them as agents to recruit miners from their own communities.[28]

The relationships between the AACo and its British miners in the 1840s were turbulent and were arguably the precursor to the struggles for control of the industry which developed in the 1850s and 1860s and which continued into the 1900s, and to an extent still exist today. Of course those relationships were also a reflection of the background and experience of the individual managers and miners, the great majority of whom had come from Britain.

Historian J W Turner captured the essence of those early years: "For its first ten years in coal mining the company had relied almost

wholly on convict labour which proved to be effective enough to make
the prospect of transferring to free labour unattractive…. In its second
decade at Newcastle, as prisoners disappeared from the ranks of its
miners, Commissioner King had to deal with immigrant labour at its
most truculent. ..The militancy of these miners was partly the result of
their experience in the British coal industry where conflict between the
colliery proprietors and their miners was already intense, but they would
also have been affected by the experience of migration. Departing from
unusually tight-knit communities, they were subjected to the boredom,
sickness and occasional danger of a long sea journey and landed in a
strange environment with an obligation to serve an unknown employer
for seven years." [29]

Coal market starts to grow

The colony experienced a depression in the 1840s, but it also saw
important developments which affected the market for coal. These were
the growth of steamships and the growth of the gas industry (which
used coal to produce the gas). The 1850s then saw the gold rushes and
the associated major expansions in population of NSW and Victoria,
and the development of railways which initially used wood to fire their
locomotive boilers, but would commence using coal in 1961. All these
factors meant a rapidly expanding demand for coal. Steamships which
provided intrastate services and regular and faster trade links between
NSW and the newly established colonies in Melbourne and Adelaide were
the major factor in the growing demand for coal in the 1830s and 1840s.
The Australian Gas Light Company (AGL) commenced operations in
1841 and became a major user of coal in the following decades. Gas
production in Melbourne also came to rely on NSW coal, as did the
Victorian railways until around 1910, when, following the disruption to
coal supplies by strikes in NSW, coal mining began in Wonthaggi.

During the 1840s AACo's monopoly was becoming less and less popular
and was under threat from other producers. A select committee inquiry
of the NSW Legislative Council into the coal industry recommended the
abolition of the monopoly in 1847, but events were already underway to

make the monopoly irrelevant. The Brown family started to mine coal at Four Mile Creek near Maitland around 1843, winning a contract to supply the Hunter Steam Navigation Company at Morpeth on the Hunter River. The Government and AACo took legal action against James Brown, the eldest son, and a case was heard in the NSW Supreme Court in 1845. Ironically, the Browns' mine was on land leased from Captain Dumaresq, the son of Lieutenant Colonel Henry Dumaresq. Henry Dumaresq had been the Commissioner of AACo in the 1830s. "Of all people they could have chosen as landlord, the son of the Australian Agricultural Company manager was surely the one least likely to acquiesce in an unauthorized coal mine on his property."[30] James Brown lost, but following an appeal, the government did not wish to see him punished and damages of only one shilling were awarded. In fact, the Attorney General stated that there was nothing to prevent mines being developed on land in the colony, not including the AACo's grant, thus making it clear that the monopoly was effectively only a prohibition on others trying to operate on land granted to AACo. AACo recognised that its monopoly had effectively ceased to exist and, following negotiations in London, the termination of the 1828 agreement was formally announced in 1847 when the British Colonial Secretary instructed the NSW Governor to end the monopoly and to free up the industry.

J&A Brown, the company which emerged from the Brown family operations, was to become a significant player in the industry. In 1853 brothers James and Alexander Brown, were commissioned to build a shaft for two business partners (Eales and Christian) for what was to become the Duckenfeld colliery at Minmi west of Newcastle. The Browns were able to obtain the mineral rights for a large area of land adjacent to Duckenfeld and began mining on this land in 1857. Eales and Christian then sold the Duckenfeld mine and land in 1859 to the Browns[31]. Eales and Christian had obtained approval to construct a railway from Minmi to Hexham from an Act of Parliament in 1854. However the Browns now wanted to extend the railway to the Hunter River and so needed the authority to cross the Newcastle to Maitland railway line. The Government amended the 1854 Act in 1861 to give the

Browns the access they needed to the River. They now had the means to freight their coal from Minmi direct to the River at Hexham where they erected a steam crane and loading shoots; they were now able to load coal into their own small ships which then took the cargo down the River to Newcastle where it was loaded onto larger ships for export overseas, or for shipping to Sydney or other Australian ports[32].

Minmi, the J&A Brown company town, developed as a significant urban centre, reaching a population of around 1,000 by 1877, with around 800 miners employed at the Brown collieries. However it was a primitive town, with crude dwellings, and was described as a "city in the desert" as at that time it had no decent road linking it to the major nearby centre of Wallsend.[33] By the 1860s J& A Brown was one of the dominant producers in NSW, and remained so right through to 1931 when the then head of the company died and the company merged with another major producer to form J&A Brown and Abermain Seaham Collieries Ltd (or JABAS as it became known). JABAS became part of the Coal and Allied group in 1960, and then part of the Rio Tinto group in the 1990s.

The start of the NSW railway system was the Sydney to Parramatta line, completed in 1855. This was followed by the construction of other lines, including the Newcastle to East Maitland line, completed in 1857. The next three decades saw the rail system expanded to many regions in NSW, so that by 1890, over 3400 kilometres of track was in use. In 1861 trials by the NSW Railways using coal for its locomotives proved successful and coal then replaced wood as the fuel for the system, opening up a major and growing market, initially for the Newcastle producers, and in due course for the western and southern producers.

With the opening of the Newcastle to East Maitland line in 1857, there was now competition between rail and the steamships which plied the Hunter River from Morpeth near Maitland to Newcastle. The Commissioner of Railways was initially unsure of how his new rail line would perform, saying that it had to compete with the steamers on the River, with no large population centres existing at the either end of the line. To be successful, he said, the line would have to depend on the traffic from the developing coal industry, and on further extension of the line into the Valley.[34] The Commissioner was too pessimistic, as the development of the coal industry in the 1860s and beyond saw the rail line used to transport coal to the port of Newcastle, with the construction of a number of branch lines to link mines to the main line. The 1854 Act to allow the construction of the Minmi to Hexham railway was followed by other Acts including an Act in 1860 to allow Newcastle Wallsend to build a line from its mine to the main line; the Morehead and Young Act of 1963 giving Scottish Australian the power to construct its rail link from Lambton colliery to the main line at Waratah; and the Waratah Act of 1863 giving the Waratah Coal Company the power to construct its mine and its railway link to the main line.

The struggle between miners and owners begins

The industry was further opened up in 1850 when a notice in the Government Gazette cancelled "reservations of coal to the Government in grants of land past and present". This meant that owners of land also owned the rights to the minerals contained in that land. That change and the abolition of the AACo monopoly opened the way for new producers to enter the industry, and freed the Browns to operate legally and expand their coal operations. The major producers to commence in the 1850s were the Newcastle Coal and Copper Company and the Newcastle Wallsend Coal Company, although the latter did not begin to produce any significant quantities until 1861. Newcastle Wallsend continued to be a major producer right through the 1900s, although it was subject to takeovers and became part of larger groups.

While there were some major conflicts between AACo and its workers in the 1840s, the 1850s arguably marked the start of the more widespread struggles between the miners and their employers which have come to characterise the history of the coal industry in Australia. The discovery of gold near Bathurst in 1851 and in Victoria in the same year led to an exodus of many miners from the coal mines to try their luck on the gold fields. The miners who remained were then in a stronger position to bargain for higher wages with their employers and did so successfully for the next few years.

One strike in 1854 involving the AACo saw the miners stop work in support of their demand for an increase in their wages. To break the strike the company brought in some miners from among those who had recently arrived in Melbourne. "The incorrigible and turbulent hands were got rid of"[35] and the company introduced a new system of payment involving a bonus for increased production. "The company threatened to dismiss any miner who got less than twelve tons per week, in contrast to the action of the union only a few years later when it imposed a maximum on its members." [36] At the time of this strike, it appears that the first "darg" or worker-imposed quota on daily production was in place at the AACo colliery. The company's superintendent later in the 1800s, Jesse Gregson, wrote that the most industrious of the AACo miners had not worked more than three and a half days a week for several months. This darg may therefore have been an unofficial arrangement among the senior miners, those who were able to earn the highest wages and therefore able to forego a full week's wages[37]. Dargs imposed by the union would later become a feature of the industry.

The Illawarra and Lithgow areas start to develop

September 27, 1849 was an historic day in the Illawarra region with the first delivery of coal from the new Albert mine at Mt Keira to be loaded onto the steamship William the Fourth. This mine was the first in the region and was developed by James Shoobert. That day saw a big turnout of locals and VIPs from Sydney and there was a procession from outside Wollongong to the port. A letter in The Sydney Morning Herald[38] reported that "a large number of the resident gentry and respectable farmers assembled at the Cross Roads leading up to Mt Keira, a distance of one mile from Wollongong, at precisely one o'clock P.M." after which they "proceeded to escort the coal carts to their destination…" The procession to the port included a band, horsemen carrying a Union Jack flag, miners, and a "goodly number" of pedestrians. Shoobert was also part of the procession. As the procession entered the town, its ranks swelled to include "numbers of Sydney gentlemen and others." Despite these celebrations to mark this historic delivery of coal to the port, and local hopes for a long future for the mine, the mine closed in 1856.

The Osborne family opened another mine at Mt Keira in 1857 which was to become known as both the Osborne-Wallsend mine and the Mt Keira mine. In the same year a mine was also developed by Thomas Hale at Bellambi, and its first shipment was made from its own jetty to Sydney in December of that year. The Osborne-Wallsend/ Mt Keira mine became the longest operating mine in Australia. It was purchased by BHP in 1937, changed its name to Kemira Colliery in 1955 and finally closed in 1991. In 1858 around 25,000 tons of coal was produced from the Osborne-Wallsend mines and the new mine developed at Bulli. The Illawarra region was now emerging to become the second major coalfield in the colony.

The first coal may have been mined or extracted in the Lithgow area as early as 1838 by Andrew Brown from his own property, Cooerwull. But the first mining on any significant scale in the area was oil shale mining. This commenced in 1865 at Hartley Vale and involved the Hartley Kerosene Oil and Paraffine Company and the Western Kerosene Oil Company. Kerosene was a popular fuel for lighting and found a ready market in the colony. The first commercial coal mine did not develop until 1868, when a syndicate of English railway workers (Poole, Woolley and Anderson) opened a mine on land owned by Rev. Colin Stewart to service workers building the new railway[39]. This mine became known as the Hermitage mine.

The completion of the Zig Zag railway linking Sydney with Lithgow led to significant development in the area. Cremin describes this time of the town's development: "Immediately after the railway was opened in 1869 there was a frantic acquisition of land in order to profit from the coal which was abundant, and, in portions of the valley, was easily worked by tunnels or shallow shafts. The first persons to use the coal were….the existing landowners. Andrew Brown had long mined coal for private use and for his mill…but he never operated commercially."[40] Another Brown – Thomas – was a major player in the development of the industry in Lithgow, developing two mines at Esk Bank in 1868 and in 1873, and supplying coal to an ironworks opened in 1874 and later to a copper smelter. By the mid-1870s there were five significant mines operating in the Lithgow area – the original Hermitage (which became part of the Bowenfels Coal Mining and Copper Smelting Company), Eskbank 1 and Eskbank 2, a new Hermitage colliery, Lithgow Valley colliery, and Vale of Clywdd colliery.[41]

Queensland's early coal history

The first discovery of coal in Queensland is generally accepted as being by Captain Logan of the Moreton Bay Settlement at Limestone near Ipswich in 1827.[42] Further discoveries were made in the ensuing years, including by botanist Allan Cunningham near Ipswich in 1828. The Moreton Bay area was opened up to free settlers in 1842 and mining began shortly after. John Williams opened a mine on the south bank of the Brisbane River near Goodna in 1843, but its life was short-lived and the mine was abandoned after flooding four years later. Williams opened his second mine in 1848 across the River at Moggill. The Hunter River Steam Navigation Company was operating a service between Newcastle and Brisbane in the late 1840s, but had to carry enough coal on the run to Brisbane for the return journey. The company threatened to stop the service, but Williams was able to secure a contract to supply coal, possibly the earliest major commercial coal contract of that era in Queensland.[43]

Queensland became a self-governing colony on 10 December 1859, winning its independence from NSW. Workers in the coal industry also began to demonstrate their independence shortly after, with the first strike recorded in the industry in Queensland occurring in 1861, just two months before the big district-wide strike in Newcastle. This was at Redbank and involved a demand for higher pay and lasted for two weeks, with 8 miners arrested for blocking access to the mine.

While a number of new mines were developed in the Ipswich area in the 1840s and 1850s, some were forced to close following major flooding in 1857 and 1858, and the industry was also held back by obstructions in the Brisbane River. With no rail link to Brisbane, the coal industry relied on the River to move its coal to market. But the River was cleared in 1866 and from that time regular steamship services between Ipswich and Brisbane were able to operate and the industry began to develop to supply the local steamship trade, the Queensland coastal steamship trade and the expanding processing industries in the Brisbane area which needed coal.

The 1860s also saw new discoveries of coal in other areas of the state, including near Rockhampton in 1862, at Laidley and Burrum in 1863, at Blair Athol in 1864 and on the Bowen River in 1867. But the coal industry was a very small part of the state economy, with employment in 1871 only 122. For the decade to 1870, the Queensland industry produced a total of only around 230,000 tons.

The 1860s was also the era when the man regarded as one of the pioneers of the coal mining industry in Queensland entered the industry.[44] This man was Lewis Thomas, a Welshman who had worked in the Welsh lead, coal and iron ore mines as a teenager and young man before migrating to Australia in 1859. Thomas tried his luck on the Victorian goldfields, but was unsuccessful, and in 1861 secured work in coal mines in the Redbank area. He was involved in the construction of a rail tunnel through the Little Liverpool Range, and then from 1866 Thomas and his partner, John Thompson, worked the Bundamba mine. Thomas opened the Aberdare Colliery at Blackstone in 1870 and then another new mine at Dinmore, where he installed coal chutes to enable efficient loading of coal onto steamers which serviced the Ipswich to Brisbane run.

The opening of the Brisbane to Ipswich railway in 1875 was an important development for the region, but the Ipswich coal mining companies would have to wait until 1885 to see the railway system extended to the South Brisbane wharves on the Brisbane River. The coal industry in the West Moreton region now had another transport option; rather than loading their coal onto vessels on the Brisbane River, provided they had the necessary access to the main line, mines were now be able to rail their coal to Brisbane more quickly and more economically. These wharves would serve as the loading facility for West Moreton coal bound for sale to other areas in Queensland, interstate, and overseas for the next 80 years.

From 1877, Thomas' business expanded and in 1881 he built a private branch line connecting his Aberdare mine to the main rail line at Bundamba. By the mid-1880s, the major complaint from Thomas and his fellow coal mine owners seems to have been the shortage of rail wagons supplied by the Queensland Railways.[45] However the availability of rail transport gave the coal industry a major stimulus in the 1880s. Production and employment state-wide surged, with most of the activity in the West Moreton. Production grew from only around 100,000 tons per year in the early 1980s to over 300,000 tons in 1890; employment was only around 250 in 1882, but by 1890 this had grown to over 800.

An article in the Brisbane Courier in 1891 gave some details about the Aberdare mine, the largest in the State.[46] The mine had been operating continuously by that time since 1866 and employed 150 miners and 50 day men (men who were employed to do non-mining jobs and were paid on a daily basis). Contrary to the general assumption that all his employees were Welsh, the article says that only about half the employees at that time were Welsh. The company owned over 20 cottages in the small town of Blackstone which were rented out to miners at a modest rent, and this area of the town was described by the newspaper as a "little Welsh village". There were two pits being mined, with the total workings covering several miles. Each pit had two means of access, with the main shaft to pit number one around 135 metres deep, and shaft to pit number two around 70 metres deep. The company had a major contract to supply one of the large shipping companies, AUSN, with between 3,000 and 6,000 tons of coal per month. The mine's output in 1890 was just over 121,000 tons, but due to the Maritime strike in 1891, that year's production was not expected to reach the 1890 level.

Thomas himself owned a grand mansion called Brynhyfryd (meaning pleasant hill), which he built in 1890 on a hill overlooking the town. The house was three stories high, with a tower and even a hydraulic lift, and was surrounded by terraced gardens. Thomas had become a very wealthy man, and had earned the name "Coal King". However when the 1890s depression hit, Thomas converted his company to a cooperative, the Aberdare Co-operative Colliery Company, which for ten years would be the entity which operated the colliery. Thomas's success is also said to have stimulated Welsh migration to Australia,[47] and he also left his mark by founding and sponsoring the Blackstone-Ipswich Cambrian Choir and the local eisteddfod movement, as well as establishing secondary school scholarships for Ipswich schools and for students back in Wales. The United Welsh Church in Blackstone was built in 1886 on land donated by Thomas and became a local centre for the mining community.

Alan Murray, in his history of the Ipswich region coal industry, paid tribute to Thomas's humanitarian work and his legacy for that key part of the State: "Thomas was also known for his generosity to miners and their families in times of hardship, injury and ill-health and for his support of miners' widows and children. In the late 19th century he was providing employment and homes for around 1,000 people. These good works by Thomas, and other owners, left a legacy of goodwill that lasted for many decades and set Ipswich apart to some degree from other mining communities, where there was often a deep mistrust and constant conflict between management and miners."[48] Thomas entered politics in 1893 as the member for the State seat of Bundamba, a position he held until 1899. In 1902 he took up a position in the Legislative Council, where he remained a Member until his death in 1913.

Regulation of mining begins

In the UK in 1842 a Royal Commission investigated the employment of women and children in mines and its findings "caused widespread public dismay at the depths of human degradation that were revealed." The Commission said that "Owners showed a critical lack of concern or responsibility for the welfare of their workers. It was common for children aged eight to be employed, but they were often younger. In mines in the east of Scotland girls as well as boys were put to work."[49] This Royal Commission quickly led to the passage of the Mines and Collieries Act in the same year, with the Act prohibiting the employment underground of women and girls and boys aged under 10. Further legislation to regulate the operation of mines in the UK and to make them safer was passed in the following decades. In 1850 the Coal Mines

Inspection Act provided for the appointment of mine inspectors and a requirement for plans of workings to be available for inspection; in 1860 the Coal Mines Regulation Act provided for general rules for operations of all mines and special rules for each mine, and raised the legal age for employment of boys to 12; and in 1872 the Coal Mines Regulation Act provided for mine managers to have certified training qualifications.

These Acts influenced the debate in Australia, although it was not until 1854 that the first piece of mining legislation was passed in NSW. The local Act was a one page piece of legislation, providing for the registration and inspection of coal mines in the Colony of NSW, with the Governor having the power to appoint any person as an "Examiner of Coal Mines". The Act also required mines to keep plans of their workings and to provide these to the Examiner. The 1854 Act was repealed by the 1862 NSW Coal Fields Regulation Act which gave the Government power to appoint inspectors and also made the employment underground of children under 13 illegal. It is not clear how many children were employed in NSW mines, but, while there were no women or girls, some boys were employed as trappers (they opened and closed wooden doors to regulate the flow of air) or as assistants to their miner fathers. Fortunately, the conditions in which such boys worked, while not good, were a far cry from the horrific conditions in Britain revealed by the UK Royal Commission in 1842.

The 1862 NSW Act prohibited the employment of children less than 13 years of age, gave the Government the power to appoint examiners and inspectors, and saw the introduction of general rules applying to the operations all collieries. The Act made it mandatory for all collieries to have two openings, required collieries to have adequate ventilation (but did not specify a volume of air to be supplied as the union had sought), required each colliery to draw up its own special rules (rules which were specific to the operation of that colliery), and required colliery management to report any deaths or serious injuries to the Minister or Examiner within 24 hours. The Act also had other provisions such as the requirement to fence shafts when not in use and to fence engine flywheels.

In 1861 the coal miners' union was fighting to have some major changes made to the coal mining Act, including a provision to limit the operating hours of collieries which would have limited the effective working day of mine workers to 8 hours. The 1862 Act was a major step forward compared with the 1854 Act, but from the mine workers' point of view it did not achieve their aim of an 8 hour day. And while the Act introduced the positions of Examiner and Inspector for the first time, for some years the effectiveness of the men in these positions would be limited, particularly when the industry's development began to surge ahead and a number of new collieries were developed.

The birth of the coal mining union movement

Despite the development of the Illawarra and Lithgow areas, the Newcastle coalfield remained by far the dominant coal mining area in NSW in the 1860s. This was the time when organised workers began to make their mark on the industry and when the employers also began to cooperate to counter the influence of the unions, and to attempt to control prices.

Of course, the Australian coal mining industry of the 1800s was in many ways a reflection of the industry in Britain from which so many of the miners and managers had come. The concept of unionism "came with British miners from Britain, and the purposes and attitudes generated in Northumberland, Durham and Fyfeshire, were as appropriate at Newcastle on the Hunter as at Newcastle-on-Tyne. Methods of working the coal, the life in mining villages, and the outlook of the management were imported directly from England. Throughout the nineteenth century the majority of union leaders had begun their working lives as pit-boys in the old country. The leading mining companies had their headquarters in Britain and the managers received their training there."[50]

The union movement's origins in the NSW coal industry can be traced to the establishment of union clubs, the first of which was set up in Newcastle in 1857 at the AACo's Borehole colliery and became known as the Borehole Lodge. It is ironic that the Borehole Lodge was formed after the AACo General Superintendent, Arthur Hodgson, drawing on his experience in Britain, urged the 300 Borehole colliery miners to form their own club.[51]   Hodgson had just arrived from Britain, and at the ceremony to mark the start of development of the company's new pit in May 1857, he addressed the gathering of the company's mine workers and managers. Hodgson said that he was sorry to learn that the miners had no club or sickness fund of their own, apart from about 30 who belonged to a branch of the Order of Odd-Fellows, and he invited them to form a new club which would be funded out of their earnings, which he reminded them were "considerable". Should sickness or misfortune strike, the miners would benefit from the fund, which Hodgson said would also assist the miners to unite "in the ties of friendship and goodwill". In the same speech, Hodgson also complimented the workers on their work ethic and behaviour, saying that the resident manager of the colliery, Mr Whyte, had told him that they were "the most industrious, well-conducted and contented set of men that he had ever had to deal with." [52] Hodgson would not be so complimentary in 1861 when the miners' union began to really flex its muscles.

The establishment of the Borehole Lodge in 1857 was followed by several other lodges in the district, including the Glebe Colliery Lodge in the same year. There was also a lodge functioning at the Minmi colliery in 1858.[53] While these early lodges were established for mutual benefit rather than industrial purposes, it would not be long before the mutual societies to which the miners now belonged provided the platform for them to develop into the precursors of trade union branches.

In May 1860, only three years after the Borehole Lodge was formed, the district miners' union, the Coal Miners' Association of Newcastle, was formed with five lodges and with James Fletcher as its first president. This was the first district mining union and it soon led to the employers seriously considering the formation of their own association. The year 1861 marked the start of much wider and coordinated action on the part of the union and the employers. In February an aggregate meeting of mine workers decided to set a standard wage for a day's work, which translated into a maximum production per man of 2.5 tons a day. This was the first 'darg' in the industry, a practice also inherited from Britain. The workers' intention was to spread the available work more equally among all the miners.[54] The meeting also resolved to support an 8 hour day for the industry, although it would wait many years for this dream to be realised.

The union held its first annual meeting in May 1861 at Waratah which drew a crowd of almost 700 and was chaired by James Fletcher. Representatives from five lodges were at the meeting – the Borehole, Glebe, Minmi, Wallsend and Tomago lodges. With a new coal mining Bill then being considered by Parliament, the meeting agreed to petition Parliament to ensure that lifting machinery was not operated more than 10 hours each day; the effect of this measure would be to limit the working day for mine workers to 8 hours. It also resolved to seek an improvement in the ventilation provisions in the legislation, and agreed to form a committee with representatives from each lodge to review the coal Bill, with a view to having the member for Northumberland (the seat which included the Newcastle area) re-introduce the Bill into Parliament. George Curless, one of the union leaders, spoke on the need to limit hours, saying that there were three reasons for shorter hours: to preserve the strength of the mine workers, to give time for recreation with families, and to enable more men to be employed.[55]

In August 1861 the owners of the four major collieries in the Newcastle area (AACo, Newcastle Wallsend, Newcastle Copper and J&A Brown) met "for the purpose of taking into consideration the propriety of combined action amongst the Coal proprietors in the neighbourhood of Newcastle with a view to check the unjust and exorbitant demands of the Miners."[56] The meeting was chaired by AACo's superintendent, Arthur Hodgson, and decided give the miners 14 days' notice of a reduction of 20% in the wage rate. The coal companies' agreement to take combined action was the first time that we see such action by the proprietors in the industry. This action did not involve the formation of any sort of industry association or legal agreement that bound the companies to act as one and its strength would soon be tested.

Shortly after the companies gave 14 days' notice of the wage reduction, the miners met and agreed to resist the attempt by the companies to lower wages. Their meeting also resolved to develop a cooperative mine which they, the miners, would own and operate. By opening their own mine they hoped to earn some of the profits being won by the coal companies, and also to reduce the industrial power of the mine owners. The mine workers then went on strike, which one local newspaper saw as a battle for control of the industry between the miners and their union and the "masters" (the owners of the mines): "The quarrel, in fact, appears to be quite as much one for power and control, as for the mere money wages which forms the ostensible cause. The masters are said to be determined to 'break' the cooperative union of the miners in an association, by provoking this 'strike,' and thus to regain the full control of their mines and men, which the combination of the miners had latterly taken from them." [57]

The strike affected all the area's mines except one and lasted for almost two months. It was broken when J&A Brown decided to come to its own arrangement with the union. The settlement saw a return to work on largely the same terms as before the strike, but this first major district dispute was hardly a win for the employers. The miners were successful in establishing their own mine, the Co-operative Colliery, which operated for eight years before it became insolvent and was taken over. James Fletcher was the driving force in the establishment of this mine and became its first manger, and he would go on to become a Member of Parliament from 1880 to his death in 1891 and would twice hold the position of Minister for Mines. Fletcher also became the owner of the Newcastle Morning Herald and Miners' Advocate.

In the early 1860s, in the midst of the industrial disputes wracking the industry, there was debate about the earnings of coal miners and whether they were being paid too much or too little. One miner's letter to the Sydney Morning Herald in 1861 gives a little insight into how he saw his job. Signing himself 'A Coal Getter' he wrote: "To show the public something of coal getting is to hew two tons of coals per day out of a solid seam two feet nine inches to four feet thick, with one hundred and fifty feet of earth on top of the seam. This is no little day's work for one man, and after he has hewed those coals he has got to fill them into skips or wagons, in a place not four feet high; that is not all; he has to prop the top up with timber, and, nine cases out of ten, the miner has a vast amount of water to contend with; the miner has to lie down in a certain position to hew the coals; and in many instances to lie down in several inches depth of water. Now any reasonable man would say it is a good day's work for any man, but his work has to commence after hewing and filling; he has got to wheel them about six hundred yards to the pit, seven or eight hundredweight at a time – that will take him six times to go back and forward with all the strength a miner has to push them along."[58]

A bitter dispute wrecks the union

A new superintendent of AACo, Edward Merewether, took over from Arthur Hodgson towards the end of the 1861 strike. The company's London bosses had approved a policy which was designed to destroy the union, which it saw as an "unwelcome novelty in its pits."[59] However, Merewether was cautioned by the company's mining expert in London that, before bringing on a fight with the workers, he should reach an understanding with the other coal companies in the district. This now set the scene for the next major battle.

The settlement, or more correctly the collapse, of the 1861 dispute did not herald a long period of peace in the industry. The following year saw an even more bitter dispute, one of the longest coal mining disputes of the 1800s. The 1862 dispute saw AACo effectively declaring war on the union, recruiting around 240 miners from the Victorian goldfields and taking them to Newcastle by ship to work in its mines. The company also recruited around 80, mostly Cornish, miners from South Australia. This was a time when the gold rush in Victoria had passed it speak, and there were many unemployed and starving miners and their families still living on the goldfields desperate for work. The miners recruited by AACo signed an agreement to work for 3 months, an agreement which included the provision that they would not join the union. They were shipped to Newcastle, where they were met by union members and urged not to help the company as strike breakers. Many of the recruits slipped out of town or refused to work for the company, and some were even prosecuted by AACo for breaking their contracts. In the end, perhaps half or less of the recruits actually went to work in Newcastle.[60] AACo's Jesse Gregson, who took over responsibility for the company's operations in 1875, later wrote that 100 of 240 miners recruited from Victoria went to work for the company "in spite of the threats and intimidation of the unionists."[61]

In 1862 some of the companies, particularly AACo, were determined to destroy the union and by the end of the year "the union was beginning to weaken under the seven months' strain."[62] The strike did not end in a formal sense, but some of the striking miners returned to work on the employers' terms, and the dispute saw the union at the district level virtually lose its influence for the rest of the decade, although the lodges continued to operate at most of the mines.

New companies Scottish and Waratah enter

The early to mid-1860s was also a time which saw two major new producers entering the industry in the Newcastle area. The Waratah Coal Company and the Scottish Australian Mining Company began to develop their mines in around 1862 and by 1864 were starting to make their presence felt. By 1865, these two newcomers were producing around 40% of the total Newcastle district's output. These two companies drove the expansion of the industry in the north during the 1860s, but as has often been the case in the industry, the result was substantial excess capacity and severe competition for sales to the major customers in Sydney and interstate. The producers would soon look to ways to control the price of coal.

Cooperation between producers to control the price of coal was a feature of the British coal industry in the 1800s. In the north east of England, a major centre of the coal industry, a system called "the vend" determined the amount of coal which could be produced by each mine and the selling price. Australia picked up that idea, with the meeting of the Newcastle company proprietors in August 1861 considering the establishment of an industry association, and also a possible agreement on prices.[63] That meeting did not agree on those proposals, but an agreement was reached in 1866.[64]

That an agreement was reached at all in 1866 was somewhat surprising considering the animosity between the bulk of the producers and the leaders of the Lambton colliery, R A Morehead and Thomas Croudace. Morehead and Croudace were not the most popular men in the coal industry in the late 1860s and 1870s; in fact they were detested by their peers. Morehead was the Australian agent (effectively the local chief) of the Scottish Australian Mining Company which operated the Lambton colliery, one of the largest collieries in Newcastle, and Croudace was the mine manager. Lambton had commenced production in 1864 and a key element of its strategy to gain a significant market for its coal was to operate as independently as possible from the other producers. Lambton was regarded as a "lone wolf" by the other producers.[65]

During the 1860s, with production rising strongly, prices were under pressure, falling by over 30% between 1860 and 1866. In November 1866 the northern producers, including Lambton, agreed to hold prices, although the agreement was soon to be breached. While this was not the first formal selling cartel arrangement (which was to come in 1872), it did mark the first time that the producers had combined to attempt to regulate the price of coal. In 1867 and 1868 there were three companies which were instrumental in the breakdown of the November 1866 agreement – Waratah, Scottish Australian and J&A Brown. J&A Brown operated the Minmi colliery west of Newcastle which it had re-purchased in late 1864 after selling the colliery in 1863. The Minmi colliery was flooded in June 1864 and the Browns were able to buy it back following the winding up of the company which had bought Minmi only a little over a year earlier.

With J&A Brown in financial difficulties following an unsuccessful mining venture in the Lithgow area, and with Minmi needing time and money to get it back into production, another viable mining development was needed urgently. As luck would have it, and possibly due to an oversight, the Scottish Australian Mining Company did not pay rent for two years on a Crown lease on one of its properties. Alexander Brown found out about this and was able to secure the lease on 270 acres of land containing valuable coal reserves. The Browns developed a new colliery there, and brazenly named it New Lambton. "Not surprisingly, the Scottish Australian Mining Company reacted with extreme hostility to this subtle act of corporate hijacking in lifting a prime property from its portfolio on a technicality. For many years afterwards, its general manager refused to be in the same room as the Browns, a circumstance which made such things as organising Vends and other coal industry matters hard to discuss."[66] J&A Brown was able to bring Minmi back into production in 1866 and the New Lambton colliery into production late in 1867, and was able to sell coal into the Melbourne market in 1868, undercutting the agreed price. With Melbourne an important market, the J&A Brown sales disrupted the 1866 agreement, and Scottish Australian and the Co-operative Colliery cut their prices in retaliation. In 1867 the Waratah Coal Company withdrew from the 1866 agreement, and in 1868 Scottish Australian did likewise, causing the other producers to take the possibly unprecedented step of writing to the Scottish Australian's directors in London, slamming the actions of "their agent in the colony" (ie Morehead) and drawing their attention "to the serious injury which must result to the coal-mining industry".[67] By 1868, the agreement was effectively dead.

In 1869 the producers were considering pushing up prices, but Morehead had other plans and was looking to the development of a new colliery on the company's property at Stockton. While negotiations were still underway between Newcastle producers, Scottish Australian reduced its prices. The AACo had not been involved in price cutting and Merewether said that Morehead was "a dishonourable fellow [who] ought to be held up to the scorn of all men."[68] Morehead's actions to withdraw from the agreement meant prices continued to fall, but he was intent on driving some of his competitors out of business.[69] Morehead had justified his actions to his directors in Britain on the basis that industry self-regulation would only be possible once untrustworthy producers had been driven out. Those directors had wanted the Australian producers to cooperate and generate reasonable profits and they became suspicious of Morehead. Nevertheless they did not overrule him, and competition continued unabated until 1872 when prices hit rock bottom. Company profits were poor or non-existent.

Morehead also was attacked by the editor of the local newspaper, the Newcastle Chronicle, which appealed to Scottish Australian's shareholders to sack him. A few weeks later the newspaper went on to slam all the producers (but in particular Morehead and his company) for their irresponsibility which had caused the miners, the companies and the whole community to suffer. The cause, the paper's editorial said, was a lack of unity between the producers, and the fact that they had allowed "one man in an almost irresponsible position to destroy their trade, and by destroying their trade to work havoc to all dependent thereon."[70]

James Fletcher, the Manager of the Co-operative Colliery and the ex-leader of the miners' union in the northern district, was a regular writer in the Chronicle under the pseudonym "Argus".[71] Historian John Turner believed that the paper's editorial may well have been influenced by Fletcher. A letter from "Argus" that appeared in the Chronicle a few days later said that the other coal company proprietors had "exhausted every legitimate means to convince Mr Morehead of the impropriety of his course of conduct" and had then proposed to him a selling arrangement "which would guarantee the Scottish Australian Company the largest share of the market." Argus also wrote about the "ruinous consequences" of Morehead's refusal to agree to maintain higher prices for Newcastle coal.

Fletcher's letter and a series of leading articles from the Newcastle Chronicle were republished in a pamphlet titled *The causes of the ruinous condition of the coal trade in the northern district of New South Wales*. Turner wrote that "it was widely believed that these articles played a significant part in the establishment of the Vend" and that, according to Alexander Brown in 1897, Fletcher "chiefly brought about a uniform selling price of coal and the state of the coal trade from 1868 to 1872 was very much worse than it is today … The proprietors, instead of living in affluence, were in beggary, and a colliery was the refuge for the destitute."[72]

So the late 1860s to early 1870s were a tough time for the workers and the producers. There was also the perennial problem of the intermittent nature of the industry, a problem which would continue until mechanisation effectively changed the way in which the industry operated after World War Two. In the Newcastle area and in the south, mining operations were geared to the arrival of ships in the port, and the need to load coal as quickly as possible to minimise the deterioration in the quality of the coal. 'Therefore every mine stopped and started as convenience dictated. Ships came in… and mines cut coal to supply them while they unloaded their inward cargoes and fitted themselves out for the return voyage. While the order was being filled, work at a mine went on feverishly. Men and boys and horses were driven to the limit over the longest possible hours."[73]

Newcastle producers combine to form the first Vend

Following the chaos of the 1860s, the miners' district union was effectively re-born in 1870 as the Coal-Miners' Mutual Protective Association of the Hunter River District. It now had a new constitution, three senior officials (president, treasurer and general secretary) and a board made up of one delegate from each of the 8 lodges. The producers were still not strongly united, although this would soon change and by the start of 1872 the northern producers were looking once again at the possibility of establishing a vend. They had considered a vend in 1870 and 1871, but proposals were rejected by Scottish Australian and AACo, with AACo only prepared to be involved if Scottish Australian was also committed.[74] However, by the end of 1872, the six major colliery companies, not including Scottish Australian, had agreed the terms of a vend. The agreement, which took effect from 1 January 1873, established the Northern Coal Sales Association, the organisation through which the vend would operate. Key provisions were higher prices for coal and a commitment by the companies to quotas on their production, with penalties for any colliery which sold in excess of its quota. The vend would last until 1880.

Prospects for industrial peace were therefore looking good, but the union launched a campaign for a 10 hour working day, culminating in a 5 week lock out in January and February 1873. The lockout ended with the union successful in negotiating a 10.5 hour day for 1873 and a 10 hour day from the beginning of 1874. Then, in December 1873, following months of negotiations, the vend producers and the miners' union struck a general agreement which contained some ground-breaking provisions, including an arbitration process for disputes and a sliding scale wage system. Disputes would now be settled by a council of arbitration consisting of employer and employee representatives and an independent umpire. The sliding scale for wages linked wages to the price of coal, with a floor set as the wage rate prevailing in mid-1872 when the selling price of high quality coal was 8 shillings (80 cents) a ton. With the Northern Coal Sales Association having already pushed up prices well above the 1872 level, workers quickly saw their own wages rise.

In March 1874, a gathering took place in Newcastle involving thousands of the locals, with the assembled crowd estimated by the Sydney Morning Herald as over 7000, and by the Maitland Mercury as 4000-5000.[75] It was certainly the largest public gathering in Newcastle up to that time, and involved miners, their families, other locals, union leaders and managers from the major mines (although Scottish Australian's Lambton colliery, not being a vend member, does not seem to have been represented at the meeting). The miners and producers were clearly happy with the state of the industry. Thomas Alnwick, a union leader from Wallsend colliery, chaired the meeting and read out the details of the December 1873 agreement, also saying that its adoption meant that "it would prevent the misery of strikes and lock-outs."[76]

The feeling of the crowd seems to have been reflected in the resolution which was moved by John Dixon, one of the miners. 'Dixon said he could tell the meeting that that was the proudest day of his existence, as a coal-miner, and a coal-miner's son…Newcastle had seen many great days, but never such a day as that one upon which they were celebrating the agreement come to between the associated masters and miners of the Hunter River district. (Applause.)…He pointed out that, whereas formerly capital and labour were bitter enemies, they had met that day upon one common platform and cordially shaken hands... He reviewed the history of the introduction of the sliding scale, which he said conferred a benefit not only upon matters and men, but upon the whole district. But the greatest thing they had ever accomplished, he said, was the arbitration clause in the agreement, which meant death to all strikes, and sending lock- outs to the winds (cheers)....The speaker concluded by moving the resolution...That we, the miners of the Hunter River District, had with pleasure the present agreement now m force between the associated masters and miners of this district".[77]

The next few years would come to be seen as a golden era for the industry. While problems and disputes did arise, the industry enjoyed a period of relative prosperity and industrial peace. Production rose steadily, with the NSW total increasing from 1.3 million tons in 1874 to almost 1.5 million tons in 1880. Employment surged from around 3300 in 1875 to just over 5000 in 1879, before falling to under 4700 in 1880. Average coal prices rose from the equivalent of 80 cents a ton in 1872 to $1.23 in 1875, before easing back to $1.10 in 1877 and then rising again to hit $1.20 in 1879. 1880 however saw prices drop to 85 cents.

The 1872 vend finally collapsed in 1880 when Scottish Australian won a lucrative contract to supply the Victorian railways and its price became publicly known. But trouble had been brewing from around 1877, when Scottish Australian opened a new mine which was able to produce at a lower cost than other mines and so undercut the agreed vend price. By 1878, most of the collieries in the north were operating outside the vend, and the market share of vend members had dropped below 50%.[78] That year saw the miners' union deciding to use its muscle to keep the vend operating, with strike action to be taken at any colliery which sold below the agreed price. In 1879 the union also attempted to enforce the vend rules on producers which were not a party to the vend. Scottish Australian's Lambton colliery was the centre of attention, with Morehead not prepared to agree to the vend conditions. Some of Lambton colliery's miners went on strike, but the colliery kept operating with a majority of its workers defying the union. With a lock-out also occurring at the Wallarah colliery, and the union under severe financial strain from funding striking or lock-out members, the union decided to impose a limit on production. The producers did not accept the union demands and in March 1880, the Northern Coal Sales Association decided to reduce the price of coal and the hewing rate to the levels prevailing before the 1872 vend agreement. The vend was effectively dead, and a period of relative peace and prosperity, which had begun to crumble a couple of years earlier, was over.

While the 1870s saw production and employment in coal mining grow quite rapidly, by 1877 the miners' union had decided that it had had seen too many new workers coming into the Newcastle district and introduced its policy of "last on first off", whereby the first workers to lose their jobs in any mine which was cutting back would be those who were most recently hired. This was a policy and practice which would be one of the longest lasting in the industry, with employers still fighting to have it abolished in the 1990s.

The battles in the 1860s and 1870s between the producers and the mining workers, and between the producers themselves, also need to be understood in the context of the cost structure of different mines in the north. With major new companies, particularly Scottish Australian, entering the industry in this period, there was a significant expansion in the industry's production capacity. But it is also important to look at the competitive position of the major mines, with the ratios of production to employment a useful indicator. In 1869 AACo's output per employee was 447 tons; J&A Brown's two collieries averaged 377 tons per man; and Newcastle Wallsend averaged 481 tons per man. Scottish Australian's output per man was 616 tons per man, 38% above AACo and 28% above Newcastle Wallsend. As the oldest producer, AACo also had "somewhat antiquated" machinery.[79] Scottish Australian also enjoyed an advantage over Newcastle Wallsend in its shorter distance to the port of Newcastle.[80] It should not be a surprise therefore that one company, with a distinct competitive advantage, would seek to use that advantage to win market share and profits. The comment by a consultant to the AACo, that its workers appeared to be well organised, and could "dictate their own demands", while "the combination of the owners, if there be any, being a mere rope of sand"[81] illustrated the difficulty for the producers in maintaining any ongoing and strong association which could match the strength and militancy of the miners.

Lithgow vend commences; Newcastle vend collapses

By 1880, the collieries in the Lithgow area were involved in an arrangement where they shared the contract to supply the NSW Railways, the major customer for western coal. The sharing arrangement involved one colliery (the Vale of Clwydd) being responsible for the contract to supply the railways, but with it and two others (Lithgow Valley and Esbank) sharing the tonnage equally, and with a fourth producer (Bowenfels) receiving a share of the revenue, as a kind of royalty.[82] This arrangement was clearly acceptable to the Railways and appears to have operated without major problems, but in 1884 the arrangement was formalised. That year the five major Lithgow collieries (Esbank, Zig Zag, Vale of Clwydd, Lithgow Valley and Hermitage) agreed to conduct their business through an organisation called Associated Lithgow Collieries. Unlike the infamous vend in the NSW northern district from 1906, the Lithgow arrangement was not the subject of an unsigned agreement. Associated Lithgow Collieries placed a public notice in the Sydney Morning Herald saying that they had agreed to associate and conduct the whole of their business through a central office, and requested their customers to only send their orders to the office. The notice also specified the standard price to be charged at the mine.[83] Associated Lithgow Collieries entered into contracts on behalf of the member collieries and controlled their production. The Lithgow producer vend operated until the late 1880s, when it appears to have ceased to effectively operate, but a vend arrangement re-commenced in 1892 with the formation of the Lithgow Coal Association.[84]

While the Northern Coal Association's vend fell apart in 1880, it was not long before a new vend was organised. The new vend began in November 1881, this time including Scottish Australian. There was a now major difference, with the producers allowed to sell as much coal as they wanted, as long as the price was above the minimum level agreed.

This vend operated smoothly and without major disputes among the producers for the next few years, but by the late 1880s, with average prices falling and non-vend producers accounting for an increasing proportion of the industry's production, the writing was on the wall. 1893 was the last year of this version of the northern vend.[85]

AACo fails to destroy the district union

AACo reduced the rate of pay for its miners in February 1882, with its superintendent Jesse Gregson hoping that this would help to break the power of the union. Gregson, arguably the leading figure amongst the employers in the late 1800s, was not anti-union, but wanted the district union replaced by a company union whose membership would be restricted to AACo employees. Gregson's stance was clear in 1881 in a letter to his superiors in London: "My objection was not therefore to Trade Unionism as a principle but to the Borehole Miners continuing to join lots with those whom they have very little in common…"[86] As expected, the company's action kicked off a strike by its miners. Gregson believed that a strike by AACo workers would force the union to break apart, and expected the strike to last 3 months, not long enough that the company could not make up for lost production in the following months. Gregson however misread the strength of the union and the strike went on for 5 months, serving to strengthen the union which funded the strikers with a levy on its members, with some financial support also coming from miners in the south and west and from other unions.[87] Under instructions from London, Gregson settled the strike.

The union then put pressure on the companies to restore the northern district agreement, with arbitration as its priority issue, wage rates at major mines having generally flowed on to other mines in the district. Negotiations dragged on over the next three years, with no real progress. In 1885, the union then began a campaign on other issues, and in particular to reduce working hours. With increasing industrial pressures on the employers, the Newcastle producers decided to take another step to strengthen their relationship and established the Newcastle Coal Owners' Mutual Protective Association in 1885. But despite the formation of this association, by the end of the year an agreement was reached between the union and the companies for an 8.5 hour day to take effect in January 1886, and an 8 hour day from July 1886. However the union would have to wait until an historic decision in 1916 before the 8 hour day was applied throughout the industry. The district union had been virtually destroyed in the 1860s as an industrial force in the Newcastle district, but the 1882 strike had cemented its role and the 1885 hours campaign showed that it continued to be a potent force.

Bulli – Australia's first major coal disaster

A major step forward in the development of mining regulation in NSW occurred in 1876 with the passage of the Coal Mines Regulation Act. The 11 page statute prohibited the employment of any female in any mine and specified that no male aged 13-18 was to be employed underground for over 10 hours during a week day, or for over 6 hours on one Saturday and over 8 hours on the next Saturday. The new Act also provided for inspectors to be appointed with the power to inspect mines at least every 8 weeks. However, with only one inspector appointed at that time, with responsibility for 27 coal mines and 1 oil shale mine, this provision in the Act was not enforced.[88]

In Queensland, coal miners presented a petition to the Legislative Assembly in 1877, demanding that there should be special regulations for coal mining, and that an experienced coal miner should be appointed as inspector of coal mines. The first inspector of mines had been appointed in 1874, but his main task was to set up and run the warden system for the goldfields. A new Mines Regulation Act in 1881 and the appointment of a new inspector of mines finally marked the real start of the era of enforcement of basic safety measures.[89] While coal mining legislation and regulation were progressing and had the potential to improve safety, real improvements in safety would be many years away.

In 1881, the first strong indications that gas in coal mines in Australia would prove to be a tragic issue began to emerge. That year, 7 miners were burnt by gas ignitions at several mines including Minmi. The mine managers and the Inspector had not seen the burns as serious enough to warrant reporting them. In the NSW Mines Department's annual report the Examiner of Coalfields said that until that year "dangerous accumulation (sic) of fire damp have been comparatively speaking unknown in our coal mines….". The Examiner went on to say that the number of men injured at different times during the year had become a serious issue, and he drew the mine managers' attention to the problem and trusted that appropriate action would be taken. The Examiner also referred to the gas problem in Great Britain and other countries where "Sudden outbursts of explosive gas, causing great loss of life and fearful suffering to those injured thereby, are the greatest dangers which have to be contended with."[90] The problems identified in 1881 were perhaps a warning of things to come.

At 2.30pm on March 23 1887 there was a huge explosion in the Bulli mine north of Wollongong which killed 81 men and boys. The Bulli mine was operated by the Bulli Mining Company. There had been many deaths in coal and other mines in NSW and the other colonies, but this was the first disaster which claimed such a massive loss of life. A coroner's inquest was held, with a jury of local residents formed as the group to decide the cause. The jury found that the disaster was attributed to wilful disregard for the colliery's own rules (the 'Special Rules' under the Coal Mines Regulation Act) and for the Act itself. The implication of this finding was that the mine management and the Mines Department were to blame. While there may well have been some bias on the part of the jury in favour of the miners, (the jury members all being local business people etc as required by the then law), the jury's findings were probably fair and reasonable.

The Government then appointed a Royal Commission to investigate the explosion. The selection of the Commissioners was questionable, with the most controversial member being the chairman, Dr James Robertson, a figure who was to play a prominent part in the Royal Commission into the 1902 Mt Kembla disaster, but more on this later. James Robertson was a mining engineer who had been involved in a major Royal Commission in 1886 which investigated the Ferndale,

Lithgow and other colliery accidents. However he was closely involved
with the coal companies in the Illawarra district, and actually appears to
have been the managing director of Mt Kembla Colliery.[91] The Illawarra
Mercury came out strongly against Robertson's appointment, stating that
he was closely linked to the Mt Kembla Coal Company and was also
believed to be a member of the Illawarra Colliery Proprietors Defence
Association, another member of which was the Bulli Mining Company.[92]
The Mercury also referred to Robertson's recent involvement on the
company side in an industrial dispute, as well as the fears of employees
for their jobs if they reported gas in the mine. Robertson was to state
fifteen years later to the Royal Commission into the Mt Kembla disaster
that he had been associated with the Mt Kembla mine for around twenty
years. The link with that mine was also important because its owner,
Ebenezer Vickery, was also a director of the Bulli Mining Company.

The other members of the Royal Commission were the managers of
the Wallsend colliery (John Neilson) and the Lambton colliery (Thomas
Croudace) in the Newcastle district, the secretary of the miners' union in
the Lithgow area, two checkweighmen (one from Mt Kembla mine, the
other from Lambton mine), and a stipendiary magistrate. The Illawarra
Mercury also raised concerns about Neilson, who had recently been
appointed by the Bulli Mining Company to do work for the colliery.
It went further by saying that no colliery manager should have been
appointed to such a Commission as he would be sitting in judgement on
his peers. Given the mines at which the checkweighmen worked and the
involvement of Robertson and Croudace, the two checkweighmen could
also be seen to have had dubious independence. The composition of
the Royal Commission has been argued by at least one historian to have
been decided so that a decision which did not damage the Bulli Mining
Company could be expected.[93]

Henry Parkes, the Premier of the day, and his cabinet ministers were
heavily involved in the coal industry. In fact seven of the ten ministers
were directors or owners of coal mines. And in the Parliament there were
also many coal company directors or owners – at least 25 in the Lower
House, and 22 in the Upper House, including Ebenezer Vickery, who was

a major coal company owner in the region and, as mentioned, also a Bulli Mining Company director.[94] Coal was big business, and the politicians and business elite of the day were major participants. This was also the case during the 1870s, when half of the Ministry were coal owners or directors (including the Premier), as were 14 of the 38 members of the Legislative Assembly, and 20 of the 68 members of the Upper House, with the NSW Governor also having an interest in the coal industry.[95]

The Royal Commission brought down its report in July 1887, blaming the miners and their supervisors for the explosion: "The commission is firmly convinced that the carelessness, want of skill, and the loose and perfunctory manner in which the principal operations in this mine were performed by the majority of the men, and countenanced by at least the overman and deputies, were immediately connected with, and led up to, the occurrence of the final catastrophe, when, by the direct negligence of probably one man, 80 other men lost their lives."[96]

New Act takes years to pass

Following the Bulli disaster and the Royal Commission, pressure to strengthen the Coal Mines Regulation Act grew, and a Bill was introduced into the NSW Parliament in 1889. The Bill aimed to raise the standard of inspection of mines, as well as increase the minimum size of pillars, improve ventilation and strengthen provisions for the use of safety lamps. While the Bill passed the Lower House it was blocked in the Upper House, and lapsed when the Government led by Henry Parkes resigned in 1891. Another Bill was introduced into the Parliament in 1893, this time containing a controversial clause which would limit the working day underground to 8 hours. This Bill was also held up, with the 8 hour clause the basis for much of the opposition in the Parliament. Alexander Brown from J&A Brown was one of the Upper House members with coal industry interests who opposed the Bill.

Following an election in 1894, the new Government appointed a Royal Commission in 1895 to review the Bill, its members including AACo's Jesse Gregson and James Curley from the miners' union. The

Commission recommended that the essential elements of the Bill should
be adopted. The Bill was finally passed in 1896, with some of its major
provisions including the registration and certification of mine managers
by a board of examiners, the continued prohibition of employment of
females, the restriction of the hours boys aged 14 to 18 could work,
requirements for the use of safety lamps under certain conditions, and
detailed rules for underground lighting. The 8 hour day clause was not
part of the new Act.

The use of safety lamps was mandated in any place in a mine in
which gas was likely to be found at dangerous levels. As we will see, this
provision proved to be a battle ground for many years between mine
management and the Department of Mines Inspectorate and failed
to prevent a number of explosions, including the Mt Kembla colliery
disaster in 1902.

Major disputes followed by the 1890s depression

The 1888 year saw another major strike on the northern coalfield which
lasted for over three months. This was a different type of dispute, one
which raised the issue of the distribution of wealth between labour and
capital. The miners appealed to the public with a "Miners' Manifesto"
published in the Sydney Morning Herald: 'Repeated attempts have
been made to obtain redress by negotiations, but all have failed, and
the miners have been forced to the conclusion that such proceedings
could only be prevented by a strike or by submitting to a lockout. Those
amassing wealth and drawing unprecedented dividends from the results
of the miners' labour evidently desire and intend to make further inroads
on the earnings of the worker toiling for his roads on the earnings of
the worker toiling for his daily pittance, and to extort from labour its
legitimate reward. We wish it to be clearly understood that in this case
we are not the aggressors. We are simply defending one of the highest
principles for which workmen can contend - a principle worthy of the
unlimited support of ourselves and the general body of the workers, and
which is summed up in the words 'the unequal distribution of wealth.'
There will be nearly six thousand workmen involved in this struggle,

and we appeal to your sense of justice to do all in your power to-aid the workers now defending their rights and yours - the inalienable rights of labour."[97]

A new agreement was struck by the union and the employers which retained the wage rates set back in 1881, but provided for an increase in rates for miners working narrow seams of coal. It also introduced a new system of arbitration involving a neutral umpire to be appointed by the Chief Justice and contained a no-strike clause. However the settlement of the 1888 dispute failed to set the scene for a new era of industrial peace. The growing political power of the union movement was soon to be reflected in the formation of the Australian Labor Party, and the onset of grim economic times, falling coal prices and unemployment were just around the corner.

The 1890s began with the first major industrial dispute in Australia involving workers and employers across several major industries. The Maritime Strike kicked off in August of 1890, initially involving coastal shipping, but soon spread to involve other transport, coal mining and shearing in NSW, Victoria, South Australia and also New Zealand. "It was the full scale struggle between capital and labour that far sighted employers had been expecting for five years. Equally it was an assertion of strength by the trade unions"[98] After industrial battles in the 1880s, the coal mining employers were itching for an opportunity to take on the union, but agreed that they wanted any action to be seen to be the union's fault. The opportunity presented itself when the union, demonstrating its solidarity with the maritime union, advised Gregson of the AACo that it was not prepared to supply coal to ships being manned by non-union labour. On receiving this advice the employers began an immediate lock-out.[99] The mine employers prevailed, with the workers in the north returning to work in November, and in the south in January 1891.

While the strike resulted in the major maritime, transport and other unions being defeated by the strength of the employers, it also saw a strengthening of the organised labour movement. The formation of the first Labour Electoral Leagues in Sydney in 1891 was the effective birth of the Australian Labor Party. That year was also saw the shearers' strike, with Barcaldine in Queensland becoming the headquarters of the strikers, and the first representative of the Labour Movement in Queensland, shearer Tommy Ryan, was elected to the state Parliament in 1892.

The Maritime Strike was also critical in the history of arbitration in Australia as it led to a Royal Commission in 1891 which looked at the causes of the dispute, and in 1892 to the passage of the NSW Trades Disputes Conciliation and Arbitration Act which provided for conciliation councils to be established in each district of NSW. These councils dealt initially with disputes, but if these were not settled, they were passed on to a council of arbitration. This Act proved to be unsuccessful, largely due to the reluctance of employers and unions to refer disputes to the new councils[100]. The 1992 Act would in time be replaced by the NSW Industrial Arbitration Act in 1901.

Australia has experienced three economic depressions since the Colony was founded – the first in the 1840s, the second in the 1890s and the third, the Great Depression, which began in 1929. The depression of the 1890s followed the collapse of a number of banks and was deeper and more prolonged than the Great Depression, with GDP falling by around 18% from 1891 to 1983 and not exceeding the 1891 level until 1899.[101] Naturally, the coal industry was hit extremely hard by the 1890s depression and the decade proved to be a very difficult one for both producers and mine workers. Coal production in NSW hit a record 4 million tons in 1891, but had fallen by around 19% by 1893; production then recovered in the next few years, reaching a new peak of 5.5 million tons in 1900. Employment in NSW peaked in 1891 at just over 11,000, but fell to less than 9400 by 1896. While employment then rose to almost 11,500 by 1900, these figures do not tell us anything about the level of intermittency which was serious during that decade and had a serious impact on local communities. "The three years beginning in 1894 were as bleak as any ever experienced by the miners. Unemployment, intermittency, and falling wages reduced the mining community to desperate poverty."[102]

After holding up in the early 1890s, coal prices in NSW then fell sharply over the rest of the decade. By 1900, the average ex mine value per ton had fallen by over 25% on the 1893 level. The Queensland coal industry was also hit by the depression, but not as seriously as the NSW industry. Employment in Queensland actually grew during the 1890s, from just over 800 in 1890 to almost 1300 in 1896, although it fell slightly in the late 1890s to just under 1250 in 1900. Queensland production in 1890 had benefited from the strikes in NSW, and it fell in 1891, but then held fairly steady for the next four years, before growing again to hit almost 500,000 tons. Coal prices in Queensland held up until 1893, but then fell steadily for the rest of the decade (and kept falling, hitting a low in 1905). The overall drop in prices between 1893 and 1900 was just over 25%.

Competition between the producers saw the NSW northern vend break up at the end of 1893, and the battle between the employers and the mine workers continued during the 1890s, with the miners' union reduced to "to a state of disunity and weakness comparable with the bad days of the 60s."[103] However the new century would see major changes in the industry, including a new selling cartel operated by the northern producers and the shipping companies and the re-emergence of the miners' union as an increasingly strong industrial force. A strong miners' union would also emerge in Queensland, and the combined NSW and Queensland unions would amalgamate early in World War One to form what became known for most of the union's life as the Miners' Federation.

Ships at South Brisbane dock 1905. NAA J3109 5/201

Wallarah Jetty Catherine Hill Bay 1894. (NSW State Archives)

2

Federation to the Great War

The industry at the turn of the century

As the 1800s drew to a close and the States were working towards the birth of the Commonwealth of Australia in 1901, the coal industry was recovering from the 1890s depression and had developed into a major part of the industrial and social landscape of the country. Coal was produced in NSW, Queensland, Victoria, Western Australia and Tasmania. The industry dominated the economies of the Newcastle region, the Illawarra region, the Lithgow area and the area around Ipswich. Coal was also produced in a number of other areas in Queensland, including the Clermont and Maryborough areas and the Darling Downs. Production in 1901 was almost 6 million tons, with around 70% of that coming out of the northern NSW coalfield.

The NSW northern coalfield employed just over 8500 workers, with 13 collieries in the north employing 300 or more workers. The largest colliery, Wallsend, employed over 700 workers. These big collieries were among the largest employers in the country. Several Newcastle collieries had workings which extended out under the sea, including the Hetton colliery which had workings 100 metres deep extending out 2.4 kilometres.[104] The NSW southern coalfield around Wollongong also had some large collieries, although not of the scale in the north; total employment was around 2300. Seven mines employed over 200 workers, with three of those employing over 300. The largest mine was Metropolitan, located at Helensburgh north of Wollongong, with almost 400 employees.[105] The western coalfield around Lithgow was much smaller again, with employment totalling only around 450. The Lithgow

Valley colliery was the largest, but employed only 84 workers.

Queensland production was around half a million tons per year, the Ipswich area accounting for three quarters of the total. The coal industry continued to be a fairly small segment of the mining industry in Queensland, with employment of around 1200, compared to around 10,000 in the gold industry. There were around 30 collieries operating, but the average annual output of only 16,000 tons per colliery indicates the relatively small scale of most Queensland operations. Production in the other states was quite small - just over 200,000 tons per year in Victoria, 118,000 tons in WA and 45,000 tons in Tasmania.

Australia relied on coal for the production of gas for use by households and businesses and to drive its railway locomotives and steamships. Coal was also used extensively in factories. Coal was an essential source of energy underpinning the growth of the economy and there was a substantial interstate trade in coal, with NSW exporting around 2 million tons a year, mainly to Victoria, SA and WA. In international terms our coal industry was relatively modest in size. Coal had driven the industrial revolution in Britain in the 1800s and the British coal industry, the second largest in the world at the time, produced over 200 million tons a year. The US was the largest producer with over 300 million tons a year, and Germany produced over 100 million tons a year (as well as around 50 million tons of brown coal or lignite).

Sydney's dependence on coal

By the 1901 census, Sydney's population had grown to around 482,000. AGL, the gas company responsible for supplying the gas for the first street lights in Sydney in 1841, supplied Sydney's gas needs, drawing on coal from the NSW coalfields, and converting it into what was known as town gas in plants at Mortlake, Neutral Bay and Manly. Gas was the source of energy for the city's street lights, and was used in homes, factories, shops and offices for lighting. Some homes in more affluent areas of the city also used gas for cooking, but in most homes cooking relied on coal and wood. However the transformation of the city through the supply of electricity to homes and factories was just around the corner.

Legislation was passed in 1896 to allow the Sydney Municipal Council to build a power station to generate electricity for the city. The new station at Pyrmont finally came on stream in 1904, with the first electric street lights in the city turned on that year. By 1900 Sydney already had an extensive network of trams which were powered by steam or electricity. From the 1890s, the electricity for the trams was generated at Rushcutters Bay. A new power station at Ultimo began operating in 1899, and this station was subsequently expanded to cater for the growing network. The Ultimo station was the first of its type to use a steam turbine which came into service in 1905, making Ultimo the largest power station at that time in Australia.[106] Parts of that power station now form part of Sydney's Powerhouse Museum. By 1905, the electric tram network servicing the city and suburbs extended for around 136 kilometres, with another 19 kilometres of lines serviced by steam trams. Newcastle also had a tramway covering around 27 kilometres of track, and Broken Hill's tramway covered around 10 kilometres. Sydney's tram network was the major user of the city's electricity and would remain so until the end of World War One.

The use of electricity in factories in Sydney began to expand in the early 1900s, but it would be some years before electric motors and other electric equipment replaced steam powered machinery and equipment. The use of steam powered equipment in factories was the norm, with coal fired boilers the source of the energy to produce the steam. The electric motors which were in use in factories in the early 1900s were largely used for lighting.

A growing coal export trade

Coal exports from NSW expanded steadily from the 1860s to the turn of the century, by which time the industry had developed modest but significant export markets in the Pacific region. In 1900 exports from NSW totalled 1.4 million tons, with Chile, the US west coast, New

Zealand, Indonesia (Java) and Hawaii (then called the Sandwich Islands) the major destinations. Many other Pacific and Asian countries also bought small quantities of our coal. Exports accounted for around 25% of NSW production as far back as 1860, although the quantity was a modest 93,000 tons. While annual tonnage figures varied significantly, by 1890 the export market had grown to 672,000 tons, equal to around 22% of NSW production.

The coal export trade developed largely as a result of the way in which the British shipping industry operated. Newcastle, as the dominant export port for coal, depended on the visits of sailing ships which were on a round trip from Britain or other parts of Europe to Australia. After off-loading their cargos in Australia, these ships preferred to sail east with the prevailing winds and return via other ports in the Pacific or the west coast of North or South America.[107] And for these ships, looking to take on a cargo from Australia, coal was an attractive option. The quality was good and the coal found a ready market if its price was competitive. The major coal companies in NSW also had strong links to Britain, with a number registered in London and under London control. In addition some London shipping company owners were also directors of NSW coal companies, and some NSW companies had also opened branches in London to facilitate exports. 'By the early years of the twentieth century…the entire foreign trade in NSW coal was arranged in, and controlled from London.'[108]

Locally owned coal producers also played an important role in the development of this export trade, with J&A Brown's John Brown one of the pioneers of the NSW trade. In the 1880s, 1890s and the early 1900s, John Brown made many overseas trips, particularly to the UK and USA, and opened company offices in London, San Francisco and in Chile. He also lived for extended periods in London in the years between 1888 and 1893 and between 1899 and 1904.[109]

At the turn of the century, the influence of the shipping industry on the interstate trade of the Australian coal industry was also substantial. In fact in the 1890s virtually all of the interstate and New Zealand coal trade was controlled by a small number of shipping companies, including

Australian United Steam Navigation Company, Howard Smith, Adelaide Steamship Company, McIlwraith McEacharn, Huddart Parker and Union Steam Ship Company.[110] The trade was carved up by these shippers, who came to be known as "The Ring". However the major shipping companies did not control the whole trade, with some companies including J&A Brown, which owned its own ships, also involved in the trade. But the near stranglehold on the interstate coal trade by these shipping companies would become central to another vend which would be the subject of major legal battles beginning in 1910.

Newcastle was a bustling mining and port city

In 1901 the city of Newcastle and its suburbs had a population of around 55,000 and was the largest urban centre in NSW outside Sydney, having grown rapidly as the coal industry expanded from a population of less than 8000 in 1861. A further 10,000 lived in the Maitland district. Newcastle was not only the centre of the dominant coal producing district in 1900, but it also accounted for the great bulk of exports. In 1900, apart from the substantial shipments to other States in Australia, exports from Newcastle went to 30 countries. The city of Newcastle and its harbour were a hive of activity. On just about any day of the year, many ships of various sizes were anchored in the harbour, at the berths loading coal or waiting to load coal. In 1900 almost 1500 ships – or around 30 a week - loaded coal in Newcastle for sale to customers in Sydney, other Australian cities and towns and to customers overseas.

On the southern side of the harbour, the wharves stretched for around 1100 metres, with equipment for handling and loading coal occupying much of this area. Bullock Island (now the suburb of Carrington), which had been developed using ballast discharged from ships calling into the port, had even more extensive facilities, with about 70% of its over 2000 metres of wharves dedicated to coal-loading. The Bullock dyke as it was known had a double branch railway line servicing it, and had electric lighting to enable night loading. Stockton was a government run facility, with wharves of almost 200 metres dedicated to coal. Not far to the south of Newcastle there was also a 300 metre long coal loading jetty at Catherine Hill Bay which serviced ships called "sixty milers" carrying coal to the port of Sydney.[111]

An article in The Newcastle Morning Herald and Miners' Advocate in 1900 gives a glimpse by a visitor into the life of the city at that time.[112] Saturday was pay day for the mine workers and at night, the town was described as teeming with thousands of miners, accompanied by their wives and children, many doing business and others just out for the evening. But the industry itself was described as being unsettled, with miners crying out for higher wages, and the owners saying that this was not warranted by the low price of coal. The fact that Newcastle was the great coal centre of the north was evident from the collieries which could be seen in all directions, with surface facilities of some old collieries which were closed also able to be seen. But the article also says that from the hills overlooking the harbour towards the town, the city and suburbs appeared to be partially hidden from view by smoke which "seemed to hover over the town." The buildings in and around the centre of the city were not "handsome", but fine residences were to be found in Waratah where businessmen had their homes.

The majority of the Newcastle population who derived their living from the mines had small houses with only a few rooms and which were built close together. The houses were not very attractive, with many seeming to be discoloured by the smoke in the air. There were also many churches in the district, including 13 in the suburb of Lambton, although this was apparently balanced by an equal number of hotels. The smelter at Cockle Creek, the largest of its kind in the country, was also in operation, processing the ore from Broken Hill and other parts of Australia.

Ipswich coalfield dominant in Queensland

Of the 30 collieries operating in Queensland at the turn of the century, the majority were in the West Moreton, in and around Ipswich, Bundamba, Goodna and other nearby towns and villages. The municipality of Ipswich, which covered the main coal producing areas in the Ipswich coalfield, had a population of just over 8600 in 1901 and was one of the main urban centres (or clusters of urban centres) outside Brisbane. Queensland's total population was only 503,000 at that time. Ipswich became a city on 1 December 1904; the city's industrial base did not simply rely on coal but included large railway workshops, a cotton mill, a woollen mill and brick manufacturing.

In 1900 there were three collieries operating on the Darling Downs, and collieries in the Burrum coalfield near Maryborough, and in the Clermont area at Blair Athol. The Burrum coalfield was the only significant coal producing area outside the West Moreton and Darling Downs at that time, with production of around 100,000 tons a year.

The Queensland Department of Mines' annual report for 1900 speaks optimistically about the developments in the industry, for example with two new collieries being developed in the Ipswich coal field, one colliery at Bundamba just coming into production, and legislation passed by the Parliament for the construction of a rail link between Callide Creek and the port of Gladstone. The report also noted that Callide coal had recently been tested on the Sydney Mail Train on the leg between Brisbane and Toowoomba and had been successful. The development of the Callide line was expected to be completed within about three years, and development of mining commenced by the end of 1901. However the optimism was misplaced, as mining at Callide did not commence until 1945 and it was not until 1953 that the rail link to the Callide coal mining operations was completed. Sadly, 1900 also saw a total of 9 men killed in Queensland coal mines, 5 of these from a gas explosion at the Torbanlea colliery near Maryborough. A Royal Commission was held to inquire into the Torbanlea disaster and also into gas problems generally on the Ipswich and Burrum coal fields. The Royal Commission made some important recommendations, including on the qualifications or experience of mine managers, on mine ventilation, and on the use of safety lamps.

Despite the optimism shown by the Department of Mines, there was a good deal of concern in the industry with the coal loading facilities in Brisbane. In 1900 the Queensland coal producers in the region west of Brisbane looked with envy at the facilities in Newcastle and were pressuring the Government to upgrade the facilities on the Brisbane River which at that time belonged to the Queensland Railways.[113] The coal producers were unable to compete with the Newcastle producers for export markets, and were also struggling to compete for some markets within Queensland. In July 1900, the Brisbane Courier reported on the concerns of the coal producers and shipping companies, with the manager of one of the collieries saying that he had told the railway authorities that the cranes currently in use "should have been blown up long ago".[114] There were also concerns about the inadequate number of trucks and poor lighting on the wharves. The paper accused the coal companies of not being sufficiently united to force the authorities to make the necessary improvements, but said that with the formation of the West Moreton Coal Owners' Association (which held its first meeting in June 1900), it expected that the industry would be able to present a united position and tell the government what action was needed to upgrade facilities.

The Association was formed by the coal companies in 1900, but not in response to industrial issues and union power as occurred in NSW. It was formed because of concerns over coal prices and the need to obtain improvements in the facilities in Brisbane.[115] At a meeting of the major colliery companies with the Queensland Minister for Railways and the Railways Commissioner in August 1900, the Minister agreed to take on board the concerns of the industry. But it would be some time before facilities in Brisbane would be regarded as adequate. A report in the Brisbane Courier in June 1901 compared the rate of loading coal onto ships in Newcastle with the rate on the Brisbane River.[116] In May of that year over 74,000 tons of coal had been shipped from Newcastle in one seven day period. In contrast, a consignment of 10,000 tons of coal from Brisbane had taken ten days to load using the primitive equipment on the wharf alongside the Brisbane River. The newspaper said that trucks drove onto the wharf where there were no facilities for loading the coal apart from primitive cane baskets. It compared Brisbane with Newcastle's "miles of wharves" and its modern hydraulic cranes and other equipment.

The coal producers were still battling the Railways Commissioner in 1908 after he had raised the demurrage charge on producers, hitting them with a much higher charge for coal which was not unloaded within a certain time once it reached its destination. The major destination was the Brisbane River wharves run by the Railways. A senior delegation including representatives from 11 collieries in the West Moreton, and led by John Hetherington (Association President from New Chum Colliery) met the Minster for Railways to protest at restrictions which they said were hampering the development of the industry. The producers were seeking approval from the Minister to build their own coal wagons or lease them from the Government on the same terms as they said the producers in NSW were able to do. Alternatively, the producers wanted to revert to the old rates of demurrage as the new system had made the "coal trade unworkable."[117] The delegation stressed to the Minister that he needed to understand that they were competing with NSW producers, and said that over the previous 12 months the Railways had shown a lack of sympathy for the industry with its "almost vindictive attitude" in raising the demurrage charge from 6 pence (5 cents) to 2 shillings and sixpence (25 cents) and also reducing the time on the wharf when no charge applied. The Railways proceeded to introduce the new arrangements in January 1909. Arguments between coal producers and government service providers, particularly railways, go back a long way and the battles by producers in 1908 would not be the last.

One colliery not involved in the 1908 meetings was the Aberdare Co-operative Colliery Company, established in 1866 near Ipswich. This colliery was still operating at the time of the West Moreton Association's formation in 1900, but would only have a few more years of life before the company was wound up and the colliery closed. In 1907 the liquidator of the company stated that its demise had been due to a number of factors: the flooding of the mine and the damage done as a result, the consequent cessation of mining, the lack of adequate supervision of mining operations, and the lack of capital required if mining was to be resumed.[118]

While the size of the coal industry in Queensland around the time

of Federation was modest, the industry nevertheless occupied an
important place in the State's economy, particularly in the Brisbane area
and in larger towns and mining centres, and also to fuel the locomotives
for Queensland's already extensive rail network. Coking coal had been
produced in the north Ipswich area from the 1860s and had found
important markets in the coke works which sold their product to local
foundries and other factories in Ipswich and Brisbane and to the metal
smelters in areas including Mt Morgan, Mt Garnet and Chillagoe.[119] There
was a growing demand for coal and coke from the State's metalliferous
mines. The Mount Morgan gold mine, around 40 kilometres south west
of Rockhampton was a major coal user, having been connected by rail to
the main railway line to Rockhampton in 1898. Coke consumption state-
wide in 1900 totalled 13,000 tons, around two thirds of that total coming
from local coke works and the balance from Newcastle.

Coal was also used to produce gas by companies including the
Brisbane Gas Company, the South Brisbane Gas and Light Company
and the Ipswich Gas Company, with the State having 14 gas works in
1900 consuming 35,000 tons of coal. The growing number of factories
in Brisbane and other centres using steam boilers were also an important
market for the local coal producers.[120] However the largest single
customer for the various collieries was the Queensland Railways, which
by the turn of the century had 4500 kilometres of track open for traffic.
And from 1910 there was a major expansion of the rail system, with the
passage of legislation that would see the network extended along the
coast to Cairns and major extensions inland. The Queensland Railways
appears to have been using around 125,000 tons of coal a year across its
network.[121]

The development of power stations in Brisbane to generate electricity
had really only begun in the 1897 when the Brisbane Tramways Company,
which had made the decision to electrify its network, built the State's first
station of any real size in Countess Street, with a capacity of 0.9MW
(expanded to 4MW by 1915). This was followed by the City Electric
Lighting Company's new power station in 1910 in William Street which
had a capacity of 1.2MW.

Wollongong and Lithgow: coal towns

Wollongong at the turn of the century was a small town; with other small population centres along the coast, including Bulli and Helensburgh, it serviced the coal mines and the local population. The population of the Central Illawarra municipality was less than 5000 in 1901. The major secondary industry in the area was the smelter at Dapto which processed ore from Broken Hill and other mines and which employed around 400 to 500 people at its peak. The smelter closed in 1905, but another smelter was built at Port Kembla and opened in 1908.

The year 1898 saw the passage of the Port Kembla Harbour Act which provided for the development of a deep water harbour in Port Kembla. Private jetties had been built between 1882 and 1890, but the decision to develop port facilities which could cater for ships in all weather was a major win for the region. In 1882 the Mt Kembla Coal and Coke Company built its own jetty at Port Kembla and a rail line linking the jetty to its Mt Kembla mine. This was followed in 1890 by the construction of another jetty by the Southern Coal Company, with a rail link to the main State Government railway line. The work for this major new development commenced in 1901 and involved the construction of two long breakwaters, wharves and other facilities. Other jetties in use for coal shipments at the turn of the century included jetties at Coalcliff and Bulli. The last jetty at Bellambi was destroyed in a storm in 1898 and was never re-built.

The Lithgow municipality had a population of around 5300 in 1901. It was an important coal producing area, particularly as a supplier to the NSW Railways and other markets, and was also an important industrial centre. Iron production in Lithgow commenced in 1975, making Lithgow the second location for iron production in Australia after Mittagong where production at the Fitz Roy iron works began in 1848. But Lithgow was the first location for steel production which commenced in the town in 1900. Steel production was to continue until 1932, when the owners of the steelworks transferred their operations to Port Kembla. In 1900 Lithgow also boasted copper smelting, tweed mills, bricks and pottery, a meat refrigeration works and oil shale mines.

Sydney also had a colliery operating in 1900, then called the Sydney Harbour Colliery. This mine commenced operating in 1897, with its surface facilities at Balmain and its workings extending in due course out under the harbor. As the Balmain colliery, it produced its last coal in 1931.

Safety standards were poor and mining technology was primitive

Coal mines at the turn of the century were not pleasant places to work in. Not only was the work hard and dirty, it was also dangerous. In the decade from 1891 to 1900, there were 158 deaths in NSW coal mines; and in the decade from 1881 to 1890, 239 deaths (including the 81 killed in the Bulli disaster of 1887). The much smaller industry in Queensland saw 32 deaths in the 1890s and 18 in the 1880s. There were also many serious non-fatal injuries every year.

But the record of Australian coal industry also needs to be considered in the context of the times and of the record of the Australian metalliferous mining industry and the coal industry in other countries. Over the period from 1880 to 1899 for example, the NSW coal industry had an average of around 2.3 deaths per year for every 1,000 men employed. In Queensland there were 27 deaths in the 1890 to 1899 period, and the death rate was around 2.5 per 1,000 employed.[122] These are grim statistics, but not too far above the record of the coal mining industry in Great Britain and Ireland for the same period, where the average was around 1.7 deaths per 1,000 employed.[123]

The death rate in the NSW metal mining industry in the 1890s was above 1 per 1,000 employed for most of the decade, and was around 1.3 in 1900.[124] The Queensland metal mining industry's death rate was also high at around 1.4 for the period from 1890 to 1900.[125] The figures for the NSW and Queensland coal mines for this period were, of course, affected by very high numbers of deaths in some years, including for example the Stockton colliery disaster in NSW in 1896 when 11 died, and the Eclipse mine in Queensland 1893 where 7 were killed. Deaths and accidents were due to a variety of factors, including explosions of gas. However the most frequent causes were falls of coal or rock from the side of tunnels or from the roof.

The typical mine was a "bord and pillar" or "pillar and stall" operation which involved driving tunnels into the coal seam, with the result being a pattern resembling a city's central business district. We can think of the city roads as the tunnels, and the office buildings as the pillars which held up the roof of the mine. In many mines the pillars were then extracted in a process called the "second workings", leaving the roof to cave in as the miners retreated (or worked back).

The industry was totally male, with some of the workers only boys and youths. Out of the total NSW industry employment of just over 10,500, 537 were boys aged under 16, of whom 364 worked underground in a variety of occupations. In 1900, there were 16 fatalities in NSW, with 5 of those being youths aged 18 or under (and one who was only 15 years of age).[126] Of course the employment of young men and boys was not unique to the coal industry at that time, as it was common for boys to leave school at age 14 or 15 to find work.

Mines generally had a large headframe located above the entrance to the mine shaft in which was installed a winding mechanism; this was used to lower miners into the mine and haul them out again at the end of their shifts, and also to haul out coal and lower supplies. In some mines the men would enter the mine by walking down a sloping tunnel (or drift) and coal was hauled out of the mine by mechanically driven endless rope conveyors. Underground, the miners drove the tunnels by hand using picks, or by boring holes in the coal and rock by hand and then using explosives to blast their way forward. The miners at the coal face dug out the rock below the coal seam before explosives were placed in the seam. Once the explosives had been detonated, the miners would then fill timber skips with the loose coal using shovels or forks.

The major haulage operations in the larger coal mines in 1900 in were carried out by men and horses. Once the skips were filled with coal, the horses hauled the skips along rail tracks to a location where they would be attached to the rope conveyor system and then hauled to the shaft and then sent to the surface, or hauled directly to the surface along drifts. In small mines the skips were hauled from the face to the surface, sometimes by men and sometimes by horses. The conveyor systems were powered by steam engines on the surface; the steam was produced in coal-fired boilers, and the steam power was often also used to power other equipment on the surface and underground including fans and pumps.   In some mines, workers would haul the skips themselves to certain areas in the mine, surely one of the toughest jobs of all in the industry.

Ventilation has always been a critical part of the operations of an underground mine and most coal mines around the year 1900 still had underground coal-fired furnaces which were designed to draw fresh air into the mine. But mechanically driven centrifugal fans were increasingly being introduced into the larger NSW mines by 1900, with, for example the Corrimal and Bellambi collieries on the NSW south coast installing them that year. In a number of other mines, furnaces would remain for many years to come, sometimes with mechanical fans also in use. Forcing sufficient fresh air into the mine workings was often a problem, with miners in some cases some cases working over 1.5 kilometres from the shaft, and air flows often blocked or reduced by mine debris, roof falls etc.[127]

Electricity was still to play an important part in coal mines, with miners commonly using tallow or acetylene lamps for their personal use, and officials using oil flame safety lamps for gas testing, and kerosene lamps or flares being used to illuminate certain areas underground. Some of the early uses of electricity from the turn of the century included telephone and signalling systems, and motors on pumps, haulage systems and coal cutting machines.[128]

There were generally no shower or bath facilities on the surface for the mine workers and most simply waited until they got home to wash off the coal dust which coated their faces and arms. If miners had a hot bath after work, the water would probably have been heated on a coal-fuelled stove, with the coal being supplied by their employer as part of their wage and allowance package (one of the early fringe benefits available to a coal miner). Fortunately the miners generally lived close to their mines, and so the wait for a bath or shower was not too great, assuming that they had running water at home.

A decade of growth begins

By 1900 the coal industry had emerged from the depression years of the mid-1890s with growing domestic and overseas markets, and for a time, improving prices. But of course it would not be all plain sailing: that has rarely been the case in this industry. A major disaster was to strike Wollongong in 1902, and a few years later in 1906 the coal companies in the northern district, in association with the major shipping companies, would embark on one of the most controversial arrangements in the industry's history. The early 1900s also saw the start of a major expansion of the industry based on the coalfields west of Newcastle in the area between Maitland and Cessnock known as the South Maitland district.

And while the industry had seen many major strikes and lockouts and had established itself as a national leader in terms of industrial disputes, one of the first major "political" strikes was to occur later in the decade. The push for nationalisation of coal mining would also become a major issue, and one which would continue throughout the period to the 1980s, although the push for nationalisation would change into a campaign for government intervention in coal marketing in the 1980s and 1990s.

The Mt Kembla Disaster

At around two in the afternoon of 31st July 1902, Australia's worst man-made disaster took place at the Mt Kembla mine near Wollongong where 261 men and boys were working underground. There were two shifts overlapping – the 7am to 3pm shift and the 9am to 5pm shift. A total of 96 men and boys lost their lives during or following a massive explosion in the mine.[129] "The mine convulsed like a mythological beat. A loud rumble came from the heart of the mountain and a red tongue of flame writhed from the mine mouth, accompanied by billows of black smoke. The terrific report of an explosion rocked the mountain, and thousands of tons of stone, earth, timber and iron were belched onto the surrounding mountain side. ...The explosion was heard 20 miles away at Helensburgh to the north and Jamberoo to the south."[130]

The explosion was later found to have been caused by a build-up of what was then called fire-damp (methane gas) in the mine which had not been detected by the miners, the management or the Department of Mines inspectors. While this was a tragedy of the worst kind for the families, the mine and all its employees and the local community, it was also important in providing an understanding of the way in which the industry functioned at the time and of the entrenched attitudes on the part of mine owners, managers and miners. Explosions were nothing new to the miners and managers of the coal industry in the Wollongong district. There had been the massive explosion at the nearby Bulli mine in 1987 when 81 were killed. An explosion had also occurred in 1898 in the Dudley mine in the Newcastle district which resulted in 15 fatalities. Explosions were also an infrequent but well known occurrence in the nineteenth century Scottish, English and Welsh coal mines from which many of the miners and managers working in Australian mines had come. Horrific explosions and loss of life also occurred in some continental European and American coal mines.

By 1902 the Mt Kembla mine had been operating for nineteen years and had not experienced any major problems with gas. Publicly it was not known as a gassy mine, although internally there had been reports of gas in the mine. The mine was owned by the Kembla Coal and Oil Company Limited whose chairman was Ebenezer Vickery who owned or had a significant say in the running of a number of the mines in the Illawarra district. Vickery was also a director of the Bulli Mining Company at the time of the explosion at the Bulli mine in 1887.

Gas and naked lamps a potent combination

The Mt Kembla disaster led to several official inquiries, but before looking at these, it is important to consider the technology used in coal mines at the time for lighting. In the 1800s and into the early 1900s, electricity had not yet made any significant impact in coal mining, apart from beginning to be used for coal cutting machinery. Miners working underground used lamps which were attached to their caps; these "naked lamps", shaped like coffee pots, used tallow, peanut oil or other oil as fuel, but gave off a good light and were popular among the miners for this reason.

Having a naked flame which was exposed to any inflammable gases in the mine made these lamps potentially highly dangerous, but the alternative at the time was the safety lamp which had a flame encased within the lamp. Safety lamps were invented at around the same time in Britain in 1815 by Sir Humphry Davy (the Davy lamp) and George Stephenson (this lamp became known as the Geordie lamp). However these safety lamps were unpopular with miners as a working lamp as they gave off a much poorer light than the tallow lamps. The poorer light restricted the efficiency with which miners could work, and so potentially reduced their wages, as payment was on the basis of the quantity of coal produced. Miners were also concerned about the poor light creating more safety hazards. And safety lamps had the drawback that, if they went out, the miner would have to borrow a fellow miner's lamp and walk back to a place in the mine where it could be re-lit, again reducing his earnings for the shift and his mate's earnings. Safety lamps were also unpopular with many mine owners and senior managers. They cost much more than the tallow lamps and production could suffer if the miner's effectiveness was impacted. Many owners and managers simply did not believe that their use was justified, except perhaps for mine inspections.

As we have seen, it was in the early 1880s that the problem of methane gas in our underground mines started to come into focus, culminating in the Bulli mine disaster in 1887. The 1890s also saw two men killed by gas at the Stockton colliery in Newcastle, followed a few days later by the death of 9 others who had entered the mine to investigate the cause of the gas leak. Following the passage of a new Coal Mines Regulation Act in 1896, the first Chief Inspector of Coal Mines was appointed in NSW in 1897. For Alfred Atkinson the management of naked lamps and gas in coal mines would be a frustrating challenge for much of his long tenure as Chief Inspector. Atkinson came with a distinguished background in the British coal industry and was one of 93 applicants for the position, including 82 from Britain.[131] In his first annual report as Chief Inspector, Atkinson referred to Metropolitan as the only colliery in the State where safety lamps were used throughout the mine's underground workings. While he recognised Metropolitan as a gassy mine, even in his first year in the job, Atkinson may have been pointing to Metropolitan as the example which all other mines should follow.[132]

The explosion in 1898 at the Dudley colliery near Newcastle was caused by a build-up of gas in the mine which also mixed with coal dust, another potentially explosive element in coal mines. Following the Dudley explosion in 1898, Atkinson had actively pressured mine managers to adopt safety lamps whenever there was a trace of gas, but was generally met with opposition.[133] He had also sought to clarify the power of his inspectors in relation to enforcing the use of safety lamps. But the Crown Solicitor's advice was not encouraging: "It would appear from this that nothing can be done by the Inspectors except to point out to the Managers in those cases where they consider safety lamps might be used."[134]

There was also the tricky question of who would pay for safety lamps. At the time of the Mt Kembla disaster an industrial case was being heard in Wollongong between the mine owners' association, the Southern Coal Owners' Association, and the district miners' union, the Illawarra Colliery Employees' Association. One of the key issues here was the demand by the miners that their pay rate (or hewing rate) be increased to offset the reduced earnings which they feared would be the consequence of using safety lamps. The mine owners on the other hand opposed any increase in the miners' rate of pay.

In the modern era, it would be unthinkable for such a dangerous technology to continue to be used in a mine, but in the early 1900s, the continued use of the old lamps was a practice accepted by both parties - the miners on the one side, and the owners and senior managers on the other.

The Mt Kembla inquiries

A coroner's inquest into the explosion established that some of the miners had died "from carbon monoxide poisoning produced by an explosion of fire-damp ignited by naked lights in use in the mine, and accelerated by a series of coal dust explosions …"[135] The inquest did not conclude that anyone was to blame for the disaster. The NSW Government then established a Royal Commission to inquire into its causes and to recommend changes to the laws and regulations relating to the operation of coal mines. Three commissioners were appointed – Charles Murray, a district court judge, Daniel Robertson, the manager of the Metropolitan mine, and David Ritchie, the general secretary of the Illawarra miners' union. Robertson was the brother of Dr. James Robertson, whose title in the report of the commission was 'consulting engineer' (but who was managing director of the Mount Kembla mine). Brotherly love however was not a factor here as Daniel Robertson proved to be no fan of his brother, with Daniel and his fellow commissioners dismissing James Robertson's theory regarding the cause of the explosion. That theory centred on the disaster being caused by a fall in the roof in the goaf area of the mine, the resulting massive blast of air then igniting coal dust in the mine. If accepted, James Robertson's theory would have meant that the company bore no responsibility for the explosion. With Metropolitan's extensive use of safety lamps, Daniel Robertson's appointment to the Commission may well have been a deliberate strategy on Atkinson's part.

The Commission reported in 1903 and concluded that the cause of the explosion was a build-up of methane gas in the goaf which was ignited by a naked flame when the air in the goaf was forced out by a fall in the roof of the goaf. This explosion also led to a series of explosions of coal dust which killed many of the miners. The explosions of methane and coal dust also generated a large quantity of carbon monoxide which caused the death of the greatest number of victims of the disaster. The Commission did not attribute blame to any one individual and acknowledged that the use of naked lamps was perfectly legal under the Coal Mines Regulation Act.[136] The Commission was however extremely critical of William Rogers, the mine manager, judging him "guilty of a grave irregularity" in not enforcing the mine's own rules regarding the weekly inspection of the mine for methane. It was also critical of the Mines Department inspectors for not ensuring that the mine abided by its rules.

The Commission concluded that the use of safety lamps would have avoided the disaster and made strong recommendations for amendments to the Act, in particular the banning of the use of naked lamps in any mine in which any level of gas was detected. It also recommended changes to the Act's provisions in relation to ventilation, restrictions on the use of explosives and the raising of the standard of education of mine managers, with a manager's certificate only to be awarded as a result of a formal course and an examination. The recommendation regarding managers' qualifications arose in the light of concerns raised by miners and the poor performance of Rogers before the Commission. Rogers had earned his manager's certificate as a result of many years of practical experience rather than any formal courses, a situation which also applied to many of the managers in the industry at that time.

But the disaster and the Royal Commission did not lead to significant changes to the Act. The problem for the Government of the day was that there was no strong support from the industry for the Commission's recommendations, with neither the owners nor the miners supporting the use of safety lamps. The Department of Mines Inspectorate, under chief inspector of coal mines, Alfred Atkinson, had moved promptly to draft amendments to the Act following the Commission's report. However, as Atkinson tactfully stated in the Department's 1902 annual report: "A bill was …prepared to further amend the Coal Mines Regulation Act of 1896 (1902), regarding the use of safety-lamps and explosives, but did not reach the introduction stage."[137]

Without a mechanism to increase the hewing rate for miners using safety lamps, the Progressive Party Government could not have expected the support of the opposition Labor Party. And with such a provision it would have had the influential coal owners offside. The result was that

"…the safety lamps bill faced certain defeat at the hands of an unholy alliance, albeit temporary, between the Labor and Liberal MPs."[138]

Mine manager becomes the target

With the Royal Commission recommendations dead in the water, at the urging of the Department of Mines, the Government decided target the Mt Kembla mine manager and to make an example of him. It appointed a judge, Charles Heydon, to conduct an inquiry and charged Rogers with being unfit to discharge his duties as mine manager "by reason of incompetency and gross negligence." Heydon ruled that Rogers was guilty of two breaches of the coal mining regulations and found him "unfit to discharge his duties by reason of gross negligence." Rogers had his manager's certificate suspended for 12 months, but this was done with regret, with Heydon saying that Rogers was "a worthy man, who, without educational advantages, had risen from a humble rank of life to an honourable position by the exercise of qualities which had secured the goodwill and the confidence of those with whom he had been brought in contact."[139]

The Royal Commission and the judicial inquiry exposed the industry's fault lines – the lack of rigorous adherence to the regulations, and in particular the irregular checking for gas by mines, the failure to report gas to managers, the disregard by some managers of the existence of gas, and the use of naked lamps in gassy mines. These faults did not only apply in the Mt Kembla mine and the Department's inspectors also came in for criticism by not strongly enforcing the regulations. As one history of the event states, Rogers was the scapegoat for the ills of the entire industry.[140]

Rogers went back to his manager's role at Mt Kembla after the 12 month suspension period, but Mt Kembla mine lost no time in adopting safety lamps for use throughout the mine. Atkinson's chief inspector's report in the Department's annual report for 1903 noted that since the explosion safety lamps had been used exclusively in the underground workings of the Mt Kembla mine.[141]

The chief inspector's efforts bear fruit

While the Coal Mines Regulation Act was not finally amended to provide for naked lamps to be outlawed in legislation until 1941, there was a steady uptake of these lamps by mines in the intervening years. Atkinson was clearly a factor in this uptake and he made good use of the Department's annual reports to detail which mines had decided to use safety lamps and which were the recalcitrant ones in the industry. Times have changed and modern departmental reports tend to be much more circumspect in their coverage of industry, but Atkinson was determined to reduce the dangers in the industry from what he saw as an avoidable major hazard. His annual reports warned mine managers, who had the statutory responsibility for the safety of their mines, in effect saying that if further disasters occurred they would be at fault.

In the Department's 1903 annual report, published before the Royal Commission's report had been presented to the Government, Atkinson devoted a section of his chief inspector's report to the Mt Kembla mine. He made no bones about his own views: "It is to be hoped that this unfortunate calamity will impress upon all colliery owners, agents, managers, officials and workmen the importance of adopting such precautionary measures as science and mining progress have provided; otherwise, as the necessary elements are present in many of the collieries for producing such disastrous explosions, this State cannot in the future look for immunity from this class of accidents." He went on to refer in particular to the use of safety lamps in collieries where fire-damp was present even in small quantities, and also to the abolition of the use of gunpowder for blasting in collieries where safety lamps are in use, and the suppression of coal dust in mines by watering.[142]

The Department's 1903 annual report also saw the Duckenfeld and Brown's No. 4 collieries in the Newcastle district put on notice following reports of gas in the case of the former, and the ignition of gas which saw a miner suffer burns in the case of the latter. In both cases management had been "asked to introduce safety-lamps, but they state that they do not consider their use necessary, and have, therefore, refused to adopt them." In addition, Andrew Sneddon, owner of the Teralba mine near

Newcastle, was singled out for taking his own time on the issue. Sneddon had been requested to adopt the use of safety-lamps following a report of gas in the mine, "but up to the present has refused to do so, although he has promised that this shall be done in the month of April, 1903."[143]

In his 1905 chief inspector's report, Atkinson acknowledged that the traditional resistance by miners to the use of safety lamps was continuing but was easing: "The prejudice against the use of safety-lamps is to a certain extent dying out as the miners become more accustomed to them." He went on to again pressure managers to impress on their officials (those carrying out mine inspections) the requirements under the Act to use safety lamps to search for fire-damp, and to make a written report of any quantity found. The 1905 report was produced in 1906 after an explosion at the Stanford Merthyr mine near Newcastle in which 6 men were killed. In his report Atkinson again targeted mine managers, highlighting a point from the Royal Commissioner's report on that explosion: "A manager should clearly understand, and likewise impress on those under his control, that every discovery of gas in any quantity must be reported in compliance with the special rules under pain of instant dismissal."[144]

In 1909 there was an 11 day stoppage by miners in part of the large Aberdare colliery near Newcastle. Gas had been found in the mine in potentially dangerous quantities, but "the prejudice of the miners against safety-lamps…caused a stoppage of work…"[145] Even in 1912 we see continuing strong resistance to the use of safety lamps. Atkinson was still the chief inspector for coal mines and reported that fire-damp had been frequently found in the South Clifton Tunnel mine near Wollongong. "Representations have been made to the manager advising the use of safety-lamps throughout the mine, but up to the present he has not considered it necessary to adopt them. Under the circumstances … I consider he is taking unnecessary and unwise risks."[146]

Mine managers of course had statutory responsibility for the operation of their mines, and the Coal Mines Regulation Act and its regulations gave managers the discretion to decide whether or not to employ safety lamps throughout their mines. While questions of cost and the attitude of miners to these lamps were also important, some managers no doubt believed that the efforts of the chief inspector impinged on their managerial and statutory discretion and, like the manager of the South Clifton Tunnel mine, were determined to protect their areas of responsibility.

After the Mt Kembla disaster, adoption of safety lamps throughout the major Illawarra mines was relatively quick. Metropolitan was the only mine using these lamps throughout the mine at the time of the disaster, but by the time of the Department's 1905 report, we see that all the large collieries in the southern district had this practice. However the NSW northern district mines (the Newcastle and South Maitland districts) were more resistant to the adoption of safety lamps, one factor no doubt being that the mines in the area were generally considered less gassy than those around Wollongong. Nevertheless, the Department's 1905 report stated that 15 large mines in the northern district had reported fire damp. Most, but not all, of these northern mines were using safety lamps in key parts of the mine such as face workings and return airways. The managers of two other mines in the district were reported as intending to introduce safety lamps in areas where fire damp had been found.[147]

The Mt Kembla explosion and the subsequent Royal Commission failed to provide the circumstances for the industry to make a leap forward in its approach to safety. However there was a noticeable change in the industry after 1902, with a steady increase in the acceptance and use of safety lamps. And, following the loss of six lives at the Stanford Merthyr mine in 1905, thankfully there were no major tragedies from explosions in NSW due to gas until 1923 when disaster struck the Bellbird colliery near Cessnock. There was a major explosion at the West Wallsend – Killingworth colliery in 1910; however as the mine has been closed for two months, the only fatality was one of the pit horses. Queensland suffered a major disaster in 1921 with the explosion at the Mt Mulligan mine, but this was attributed to poor work practices and supervision.

The South Maitland coalfield starts to boom

The early mining in the Hunter region took place in and around Newcastle, but from around 1902 onwards there was a surge of development further west in the areas around Maitland and Cessnock. This development was to tap the seams of coal known as the Greta seams and this area would become the dominant coal mining district in a few short years. There had been small scale mining of the Greta coal seams from the 1860s, but it was not until 1886 that the geologist Tarrant Edgeworth David commenced detailed assessments of the seams. Results of some of Edgeworth David's work were referenced in the Mines Department's 1888 annual report which said that "geological evidence proves that the Greta measures will be found to extend …from near Ellalong to Anvil Creek, near Greta and Black Creek near Branxton."[148]

It is probably fair to say that what became known as the South Maitland coalfield, the area stretching from Maitland south west and including Kurri Kurri and Cessnock, was the Bowen Basin of its day, in the period from the early 1900s to World War Two. The Bowen Basin has been the premier coking coal area in the world for the last few decades. The coal from the Greta seams was high quality coal, up to almost ten metres in thickness in places, and was primarily used for making gas.

The East Greta Company was formed in 1891 to mine the Greta seam, and there were also other developments in the 1880s and 1890s, but it was not until the early 1900s that a number of major mines, owned by substantial companies, began to be developed to mine the Greta seams. These were to include some of the largest and most famous mines in the first half of the century, with some continuing to operate into the post-World War Two era. The NSW Department of Mines annual report for 1902 noted that developments on the Greta seam then underway included the Heddon Greta (owned by Heddon Greta Coal Co Ltd), Stanford Merthyr (East Greta Coal-mining Co Ltd), Pelaw Main (J&W Brown)[149] and Hebburn (AACo) mines. By 1910 the major mines in the area also included the Aberdare and Aberdare Extended mines (Caledonian Coal Mining Company) and the Abermain mine (Abermain Colliery Company).

These mines were big operations, by 1910 employing 800 or more in the case of Aberdare and Pelaw Main, over 700 in the case of Hebburn, over 600 in the case of Abermain, and over 500 in the case of Stanford Merthyr and East Greta. They were also far more mechanised than the mines in the Newcastle district. Electric or compressed air driven coal cutting machines (which were used to undercut the coal seam prior to blasting, as opposed to the old method of hand digging by the miner) were used extensively in these collieries. Electric lighting was also extensively used, both on the surface and underground. By 1904 there were 25 coal cutting machines in use in NSW collieries and this jumped to 59 in 1905, producing around 1 million tons of coal. By 1910 there were 181 machines in use producing around 2.2 million tons. Mechanisation was spreading throughout the industry. The South Maitland field was producing 2.7 million tons by 1910, half of the total coal coming out of the northern district. By 1915 this had grown to 3.5 million tons (56.5% of the district) and by 1920 to 4.3 million tons (63%).[150]

The Newcastle mines experience tough times

As the Greta seam mines developed, the established mines in the Newcastle area experienced very difficult times, with many collieries working only intermittently. Some of these collieries were old established operations, and were facing the last years of their lives. A union delegation representing the Newcastle collieries met the Premier in November 1904 to explain the dire circumstances faced by the workers. The union secretary James Curley led the delegation and provided information to the Premier about the operations of a number of collieries that year. The general picture was grim. The Co-Operative Colliery had operated for only 100 days up to the end of October, or around 4.5 days per fortnight; Waratah Colliery for 112 days; West Wallsend for 54.5 days; Lambton B for 26 days since late June; Burwood for 60 days for the previous six months; and Killingworth for 22.5 days for the previous 16 weeks. The delegate from Killingworth said that the colliery had operated for only 4 days during the previous 8 weeks. There had also been many workers dismissed from the Newcastle collieries, including about 300 who lost their jobs when one of the J&A Brown collieries at Minmi had closed.[151]

One delegate told the Premier that the coal producers were saying that their markets were being "filched" by the Greta seam producers who were opening new mines and installing coal cutting equipment, with some of those producers also developing new mines in the South Maitland district. The delegation urged the Premier to cease issuing new coal leases and to control the price of coal, and also to establish a local branch of the Arbitration Court, as the current system was too slow to resolve disputes. The Premier agreed to consider the miners' proposal for a local branch of the Arbitration Court for the Newcastle district, but dismissed any consideration of nationalising the industry or of controlling coal prices. The control of coal prices would happen the following year, but this would not be through government regulation, but through a new cartel arrangement involving the northern producers and the shipping companies.

The union's complaint that the producers, who were railing against the loss of markets to the new South Maitland collieries, were also involved in developing those Greta seam mines was in part correct. However the Greta seam was offering a whole new potential for existing and new companies and any producer who decided against investing in this new coalfield was throwing away the opportunity to take part in the major new field of the early 1900s. The union's desire that the government should control the development of the industry or nationalise it would be a theme that would arise time and time again throughout the 1900s.

Two of the largest and longest running collieries which were developed in the early 1900s on the South Maitland coalfield were Pelaw Main and Richmond Main, owned by J&A Brown. J&A Brown was headed by John Brown, one of the best known coal company proprietors of the era. John Brown was born in 1851 at Four Mile Creek near Maitland, where the Brown family operated a coal mine. The family moved to Minmi in 1859 and John commenced work at the age of 14, initially working underground, and then progressing to colliery clerk, surveyor and manager of the Minmi mine. In 1890, John and his brothers, Stephen and William, became the partners who managed the J&A Brown business, but John was the dominant partner. J&A Brown's operations in the 1890s were centred on the Minmi area around 19 kilometres west of the Newcastle business district. At that time Minmi was a town of around 3,000 people, owned by the company.

The Pelaw Main story began in 1897 when J&A Brown purchased the Richmond Vale estate, a few kilometres from Kurri Kurri. The company then purchased the Stanford Greta Colliery No. 2 Tunnel (part of the financially troubled Stanford Greta Colliery company operations) and associated land in 1900 and renamed it Pelaw Main in 1901. Pelaw Main and Richmond Main were adjoining properties and their location would provide valuable synergies in terms of rail transport and the provision of power. Coal was produced from Pelaw Main in November 1901 and coal cutting machines were introduced in 1902. By the end of 1903 14 of these machines were in use in the colliery and 35 were in use by 1912.[152] By 1912 the colliery employed 815 people, 606 of these underground, and that year produced 600,000 tons of coal.

Pelaw Main however had a poor industrial relations record and was plagued by strikes in the early years. It suffered one of the longest strikes ever to involve a single colliery in 1914-15, when it was closed for 42 weeks. This strike was caused by a dispute over the abolition of the second shift which ran from 3pm to around 11pm. In 1915 John Brown spoke at the company's annual picnic at Paterson, north of Maitland, and expressed his frustration with Pelaw Main. He said that the number of disputes at Pelaw Main had disgusted him, and that when a deputation of miners from the district had told him that there would be trouble at Richmond Main, he said: "There won't be. I will shut it up and it won't work a day if I have the same trouble as I have at Pelaw Main. We own the surface; we own the coal, and we can do what we like. The employees I believe, can be bright and prosperous, but not until we can supply the demands of the trade."[153]

When the company purchased Richmond Vale in 1897, there had been no coal mining since the colliery was closed in 1982, after having produced a small quantity in 1891. J&A Brown did not re-commence mining at Richmond Vale for several years, but in 1900 the company applied to the NSW Government for permission to build a rail line from the Richmond Vale Colliery to the company's rail line from Minmi to Hexham. The NSW Richmond Vale Railway Act of 1900 provided the legal approval for the new rail line. Pelaw Main initially railed its coal on another line, but it was connected to the Richmond Vale railway in June 1905 and the main rail line to Richmond Vale Colliery was completed in March 1906. J&A Brown now had full control of their transport and was not reliant on another company. Richmond Vale was renamed Richmond Main in 1911 and J&A Brown sank a second shaft that year as part of its plans to develop the colliery into a major coal mine. However the going proved difficult and although the colliery employed 283 men in 1915, no coal was produced. It was not until 1917 that coal in commercial quantities began to be produced, and by 1918 the colliery was employing 458 men.

In the early 1900s electricity was starting to become a major source of energy for mines and factories, replacing energy from steam. J&A Brown was looking to build its own power station to supply electricity to the two adjoining mines. Against the advice of his technical staff who wanted to build the plant at Hexham on the Hunter River, John Brown decided that the plant would be built on his own colliery land which also had a source of water. By 1913 part of the plant was up and running, with its power being used to drive the main ventilating fans, air compressors (for pumps and coal cutting machinery) and haulage systems at Pelaw Main and the main winding engine at Richmond Main.[154] The Stanford Methyr and North Bulli collieries were also starting to develop their own power stations that year, and a number of other large collieries had already developed power plants. The Richmond Main power station was the biggest on the northern coalfield, and in fact the biggest on any of the coalfields at that time, with one 2,000 KW plant and one 750 KW plant. The next biggest was at Aberdare Extended colliery, a plant of 800 KW capacity.

Richmond Main was the mine of which John Brown was most proud. One of the company's employees said that "It was (John Brown's) showplace; at the pit bottom he would entertain guests there – an unusual mine setting with its series of underground archways."[155] Jim Comerford, a miner who began his career in the late 1920s and later became a senior Miners' Federation leader, also attested to the mine's attractiveness and its efficiency. "If ever there was a beautiful mine, Richmond Main was it. It was so well planned, aesthetically it was pleasing. He had a world-record ninety percent extraction. The others (the other collieries in the district) barely got thirty per cent extraction".[156]

One of the often forgotten aspects of pre-World War Two coal mining was the extensive use of horses underground; Richmond Main and Pelaw Main were no exception, using hundreds of pit horses at any one time. Richmond Main's practice was to take their horses underground on Mondays and lift them out again on Fridays. The horses had their own stables underground and were looked after by an ostler, whose job was to ensure that they were properly fed and shod and otherwise in good condition. During extended strikes, horses from the Brown collieries were generally taken to the surface and let loose on the company's land, a pleasant respite no doubt from the darkness and drudgery underground.

The large collieries like Pelaw Main and Richmond Main were only partly mechanised. Many of the functions underground, including boring holes in the coal face for explosives, loading coal onto conveyors, and the widespread use of locomotives for haulage, would have to wait until the 1930s, 1940s and beyond to be classed as modern and mechanised. BHP would be the leader in this process in the 1930s when it became one of the major producers in NSW. Pelaw Main continued to produce coal until it was closed in 1961; Richmond Main continued until 1967. These two giants were part of an era which covered many ups and downs in the industry, mostly downs, and which saw massive structural change in the northern district, job losses, mine closures and the end of the South Maitland district as a significant mining area in the 1950s and 1960s.

Queensland mechanisation begins

The first indication of mechanisation at the coal face in the Queensland coal mines was the introduction of coal cutting machines into the Box

Flat colliery near Ipswich in 1905.[157] The machines were manufactured by Westinghouse of the USA and were electrically driven, with the electric power supply coming from the surface. The underground workings of the mine were also lit by electricity. An article in the Queensland Times of Ipswich said that the machines had revolutionised the operations of the mine, and gives an interesting picture of how the mine operated at that time, and of the impact of mechanisation: "Where the workmen were previously hewers or miners, they have now become wheelers and fillers. At present four rooms are being worked. The electric cutters, worked by an overseer and three men, go into each of the rooms and undermine the coal to a depth of 405 (feet). The fuses are then put in, and, after explosion, the lump is loosened. Then men come along, break the coal up, and load it into wagons. There are four fillers and one 'miner' to each room, the work of miner being to make a vertical channel and trim the edges of the seam prior to blasting operations. The man who formerly acted as check-weighman at the pit-head, finding his occupation gone, became the contractor for wheeling the coal along the workings to the bottom of the shaft. The wheeling is done by a little tafy pony that found itself introduced to these strange surroundings about a fortnight ago. It was tucked up in a net, placed under the cage, and let down the main shaft, a distance of 4051t. A stable has been made for it in the workings, and in future it will 'live' in the mine and do the wheeling. One can imagine something of its loneliness, especially during the night time."[158]

The use of coal cutting machines in the Ipswich area in Box Flat Colliery was followed by a number of other collieries including Blackheath in 1906, and the new Aberdare colliery in 1909. Greater use of electricity was also a feature of some of the Queensland collieries in early 1900s to supply power to the coal cutters, and to light the working areas of mine. By 1911, the Queensland Mines Department was reporting that 22 coal cutting machines, all electrically powered, were in use in mines in the Ipswich district, with two collieries (Rhondda and Noble Vale) having installed them that year. The Department also flagged the need to develop regulations to guide mine managers and workers in the use of these machines and in the use of electric power. The same year saw what

was described as a modern coal cleaning and screening plant installed at
the Noble Vale No.1 colliery.[159]

Industrial relations enters a new era

The early years after Federation saw historic changes in the industrial
relations systems in Australia. The evolution of the systems governing
industrial relations in the coal industry of course goes back to the 1800s
and has been briefly touched on in Chapter 1. But some of the major
changes started to occur from 1901, with the NSW Industrial Arbitration
Act of 1901, the Commonwealth Conciliation and Arbitration Act of
1904, the NSW Industrial Disputes Act of 1908, and the Queensland
Wages Board Act of 1908. The Shops and Factories Act of 1896 in
Victoria was also important in influencing the developments in NSW
and Queensland.

The first major piece of legislation to herald the new era was the NSW
Industrial Arbitration Act of 1901 which provided for arbitration of
disputes by a Court of Arbitration, comprising a judge and two assessors,
with strikes and lockouts prohibited during the period of the hearing of
the dispute. The Act attempted to provide an alternative to the disputes
which in the past had taken place after talks between employers and
unions had broken down. However this major new legislation proved
to be a failure, and by 1905 the Court was effectively paralysed following
the overturning of two key award decisions by the High Court.

Initially the various coal industry employer groups and unions
registered with the court. The employer groups included the northern
proprietors' association, the southern association, the western
association, the association representing producers and railway owners in
the Maitland and Cessnock areas, and two large producers (Wallarah Coal
Company and AACo). The unions included the employees' associations
from the northern, southern and western districts. The court conducted
detailed hearings on an award for the coal industry in the southern and
western districts and made awards for both. For both awards, the court
granted many of the miners' claims and included into the awards the "last

on first off" principle, a central tenet of the union for many years. This principle was called "last to come first to go" at the time and required the most recently hired workers to be the first to be sacked if a colliery was reducing numbers. It continued to be a cause for which the Miners' Federation continued to fight well into the 1990s.

Legal cases regarding these awards followed, culminating in High Court rulings that severely restricted the Arbitration Court's powers and rendered the award decisions null and void. Following the High Court decisions, the southern and northern miners associations both walked away from the Arbitration Court system, with the southern union negotiating an agreement directly with employers.

The early 1900s was a period which also saw other legislation passed or proposed which had a major impact on the NSW coal industry. The Miners' Accident Relief Act was passed in 1900 giving workers and families access to financial assistance in the event of death or injury, with contributions to the relief fund to come from miners, producers and the government. In 1900 the Coal Mines Regulation Act of 1896 was amended to require that to become fully accredited miners had now to work at the coal face for two years. In 1902 there were also a number of Bills introduced into Parliament which, while they failed to pass, had the potential to impact on the coal industry.[160]

These Acts and Bills and the chaotic processes surrounding the Industrial Arbitration Act caused consternation amongst the producers and in 1905 they combined to form the first state-wide association, the NSW Colliery Proprietors' Protective Association. This Association had representatives from the three districts and its purpose was "to watch the interests of Colliery Proprietors of the State in respect to Legislation."[161]

Wages boards were introduced in Victoria in 1896, and provided the example for both NSW and Queensland to introduce legislation to set up a similar structure. There was a change of government in NSW in 1907, with Charles Wade the new Premier. Wade's government passed the Industrial Disputes Act in 1908 which provided for wages boards to be established for various industries, with each board comprising employer, employee representatives and an independent chairman.

Controversially, the Act did not require the involvement of unions, and there was no provision for appeal from the decisions of the boards. There were also provisions in the Act for severe penalties for striking workers. This Act, and harsh amendments passed in 1910, would be major factors in the northern district dispute in 1909 and 1910 which would cause massive disruption to the industry. The effect of the Act "was to convince militant unionists, and indeed many moderates, that the government had declared itself openly on the side of the employers and against the unions."[162] And the fact that Premier Wade had also been the legal representative of the coal producers in various industrial cases before the Arbitration Court gave the unions another reason to be bitter towards the new government.

The Kidston Government in Queensland also passed new industrial legislation in 1908, the Wages Board Act, providing for boards to be established for various industries. For the coal industry, the Queensland legislation worked reasonably well initially, with the board for the coal industry in the south east of the State handing down the first award for the industry in 1910, mandating standard working hours and minimum wage rates for a range of occupations. The standard hours were 7am to 3.30pm Monday to Friday (including a half hour meal break), and 7am to 1pm on Saturday. The loading for overtime was set at 25% for the first 2 hours, and 50% for hours over 2. A major problem with the 1908 Queensland Act however was that it lacked a mechanism for the settlement of disputes.[163] Kidston was forced to resign in 1911 and a new government passed the Industrial Peace Act, which introduced compulsory arbitration and established an industrial court which had jurisdiction over the wages boards which were now called industrial boards. The new boards had employer and employee representation, but the employees had to be bona fide employees of an industry. Unions were not recognised in the Act.

The election of the Ryan Labor Government in 1915 saw the 1912 Act repealed, and new legislation passed in 1915 and 1916 providing legal protection for unions and a system of compulsory arbitration. The 1916 Industrial Arbitration Act, which was assented to in December 1916, cemented the role of unions in the arbitration system, recognising them as the sole representative of employees. The Act established a Court of Industrial Arbitration which had wide ranging powers, including powers to settle disputes, make awards and to provide preference to union members in employment. The major provisions of the Act would apply to industrial relations during the extended period of Labor government in Queensland.

Strikes become political

In 1905 the Industrial Workers of the World (IWW) was established in the USA to oppose the peak union body, the American Federation of Labor, and came to be a significant influence on the industrial scene in Australia. The revolutionary aim of the IWW was to unite workers in One Big Union and to overthrow the employing class. Peter Bowling, born in Scotland and with a coal miner father, became the treasurer of the miners' union in the NSW northern district in 1904 and its president in 1906. He was also a key player in the formation of the Coal and Shale Employees' Federation in 1908, which brought together the three NSW district unions under one state organisation. Bowling had been a radical for some years as a member of the Australian Socialist League. The preamble to the constitution of the IWW was a call to arms to the working class and, under Bowling's leadership, was adopted by the Miners' Federation in NSW in 1907: "The working class and the employing class have nothing in common…Between these two classes a struggle must go on until all the toilers come together on the political, as well as on the industrial filed, and take and hold that which they produce by their labour through an economic organization of the working class, without affiliation with any political party. …The trade unions aid the employing class to mislead the workers into a belief that the working class have interests in common with their employers. These sad conditions can be changed …only by an organisation formed in such a way that all its members in any one industry, or in all industries, if necessary, cease work whenever a strike or lock-out is on in any department thereof, thus making an injury to one an injury to all. "

In 1909 a disastrous strike hit the NSW coalfields. The strike was short-lived in the south and west, but lasted for four months in the northern district and led to major long term damage to the industry, including the loss of the valuable Victorian railway contracts, with the Victorian government moving to commence mining at Wonthaggi to supply its railways and with overseas markets also badly affected. Previous disputes in the industry had largely been related to wages and conditions, but in the case of the 1909 dispute the causes, while ostensibly about wages and conditions, were also more political in that the strike was also based on the belief that direct action rather than arbitration was the best option. The strike was due to a combination of factors, not least the belief on the part of the union leaders and many miners that a major dispute with the employers would force them to hand major gains to workers, thus demonstrating the benefits of direct action . An accumulation of grievances on the part of the miners, and a decision by the employers to give nothing to the miners and dig in for a fight to the finish were also key factors.[164]

During the strike, NSW Premier George Wade pushed amendments to the Industrial Disputes Act through Parliament which made it illegal for anyone to initiate, promote or assist in the continuation of a strike in the coal industry or other key sectors. Bowling, as leader of the strike, was sentenced to two and a half years in prison and was taken to Goulburn gaol in leg irons. He was scathing of future Prime Minister Billy Hughes: "The name of Hughes should stink in the nostrils of every honest worker. His deeds in the strike were the deeds of Judas Iscariot and when I go to jail tomorrow, it will be because of the treachery of William Morris Hughes."[165] Hughes, then head of the Waterside Workers' Federation, and Bowling had initially cooperated in the planning of the action against the employers. But it was not long before Hughes shifted his stance and distanced himself from the miners' union.

The miners in the western district were back at work by Christmas of 1909, and in the south by February 1910; the miners in the north did not return to work until March 1910. No gains in wages or conditions were made by the miners in any of the districts. However the strike and the gaoling of Bowling were factors in the defeat of the Deakin Federal government in April of that year. Billy Hughes became Attorney General in the new Fisher Federal Labor Government, and was now perfectly placed to take on the coal and shipping companies' selling cartel. And later in the year Labor won office for the first time in NSW following an election campaign that included posters of Bowling in leg irons. Once the McGowan Labor Government assumed office in NSW Bowling was promptly released, but his support in the north declined and he lost the presidency of the union's northern district. Bowling was secretary of the southern district union for a time, but never again was the driving force in the Miners' Federation.

The northern district was never the same after the strike of 1909-10. Coal exports from Newcastle were hard hit, and the sailing ships which were a feature of the port for so many years ceased calling to pick up coal cargoes. NSW exports, most of which were from the north, peaked in 1908 at almost 3.4 million tons, but fell to around 2.2 million tons in 1909 and 1910. Exports did recover to another pre-war peak of almost 3 million tons in 1912, before war controls and other factors had reduced the level to only 725,000 tons by 1918.

Victoria's railways, dependent on NSW coal to power its locomotives, had had enough, and the Victorian Government moved quickly to develop coal mining in the Wonthaggi area south east of Melbourne to supply the railways. The NSW contract to supply the Victorian railways was not renewed and coal production in the north fell by 2 million tons in 1909, and took another two years to recover to the pre-strike levels. The official Mines Department figures for employment show only a relatively small decline in employment in the north (less than 500 between 1908 and 1910), but these figures may hide the increase in intermittency in the industry. The Miners' Federation, which had only been formed in 1908 as the Coal and Shale Employees' Federation (the federated body of the three NSW districts), was shattered and continued in name only for the next few years, and the district unions were forced to concentrate their activities on district issues.

Queensland miners' union finally gets set

While the origins of the Miners' Federation in NSW date back to 1860, the industry in Queensland took many years to see the major mine workers union cemented into its structure. The Coal Miners' Preservative Association was established in the Ipswich district in 1878, with the association also setting up an accident fund for miners.[166] The district union continued to operate at times in the 1880s and 1890s, but it failed to attract significant membership or to be an industrial force in an industry which was dominated by small, often family owned mines.[167] The frustrating experience of the of the miners' leaders in attempting to establish a strong union was evident in 1891 when a mass meeting of miners at Bundamba was held under the banner of the West Moreton Coal-Miners' Union and was addressed by the secretary of the union. The secretary spoke of past attempts to form a union and to get the "pure union principle instilled into the workmen". But when a crisis arose in the industry, he said that all these attempts "turned out failures".[168]

In 1900, a meeting of miners in Bundamba was again debating the merits of forming a union, with the meeting chairman saying that it was time for the miners to be united. He pointed to the employers already acting as a united group and referred to recent attempts to fix a minimum price for coal which he believed would have been more successful if the workers had been united as this would have strengthened the ability of the producers to obtain better prices. While better prices of course would have provided a stronger case for higher wages, it is interesting that the chairman stressed that the proposed union should not aim to cause any friction with the producers. The major priority proposed by the chairman related to the compulsory weighing of coal - a contentious issue in the industry in both NSW and Queensland, but a critical one because the wages of many workers was directly related to the weight of coal produced.[169] While the meeting resolved to push ahead with the formation of a union, the fact that only 60 miners were present was a sign that the district was still not ready to give its strong support. In 1902 the union was effectively declared defunct. At a meeting of the union in April, where only 5 attended, including three office holders, it was agreed that the union's books should be handed over to trustees for safekeeping until such time as the union could be resuscitated.[170]

The union was revived in 1903, before fading away once more. It was not until 1906 that we see the union in Queensland being established on a strong and permanent basis. That year saw the West Moreton District Coal Miners' Union formed following a proposal from the coal producers to reduce the wage rate for the extraction of pillar coal. A meeting of mine workers appointed delegates to hold talks with the employers and also decided to form a permanent union.[171] By 1908 the membership of the union had grown to over 1,000 drawn from 20 collieries in the Ipswich and surrounding districts.[172] That year also saw the union change its name to the Queensland Colliery Employees' Union and, as already noted, the passage of the new industrial legislation (the Wages Board Act) by the Kidston Government.

The Vend: the infamous northern district coal selling cartel

In 1906, the major NSW northern district producers, with the exception of Newcastle Wallsend, entered into an agreement with the four major shipping companies to form a selling cartel. When reference is now made to "the vend "in the NSW coal industry, this is the arrangement which writers typically refer to. The previous vends or selling cartels in the north had broken down after a period of time due to over production, competition between companies and other factors.

The northern vend of the early 1900s however was a different beast to those also operating in the southern and western districts of NSW. Involving the largest coal producing district in the country and the major shipping companies, two of which were also major shareholders in large northern coal producers, it was a significant combine with a powerful influence on and control over the national markets for coal. The major shipping companies had formed a Coal Importers Association in 1897 to bid for the large coal contract with the Victorian Railways. In due course this shipping organisation became known as "the ring". It won the Victorian Railways contract and its members were allocated shares ranging from 10% to 22.5% of the tonnage involved. In 1902 the four major shipping companies (Adelaide Steamship, Howard Smith, McIlwraith McEacharn and Huddart Parker) also agreed to share the whole coal trade supplying Victoria, South Australia and Western Australia.[173]

The 1902 agreement was put to the test in 1903 when J&A Brown tried to win the contract to supply the Melbourne Gas Company, one of the largest coal contracts in the country. J&A Brown owned coal ships and had spurned offers to join the ring, but although their tender was the cheapest, they were denied the contract: "…the ring had already taken care that one of its members would gain it. When the tenders were opened by the Gas Company in the presence of the ring representatives, it was found that Brown's was the lowest; the ring promptly organised a lower price from a firm under their control, the latter's tender being accepted in preference to that of Brown."[174]

The companies in the northern district had seen strong competition in the early 1900s as the developments on the Greta seams got underway and overall production rose. Average mine mouth prices in NSW hit a low of around $0.51 per ton in 1895 during the depression years and then recovered to $0.74 by 1902. But by 1905 the average had fallen to around $0.60. There had also been pressure on the northern producers to enter into a selling agreement by the miners' union. The miners' union saw a vend as a means of stabilising the industry and raising miners' earnings. James Curley, the northern district secretary, had proposed to AACo in 1903 that "some new body newly constituted if necessary, to replace the old body if necessary, be authorized on behalf of the proprietors to fix the selling price for each year after this current year so as to determine the district hewing rate for a like period."[175]

By the 1890s and early 1900s the shipping companies had begun to own a stake in the coal producers in NSW, a stake which would become much more significant in the ensuing years. The Adelaide Steamship Company was the first to move into the Newcastle coal mining industry as the major shareholder in Abermain Colliery Company and

Seaham Coal Company. Howard Smith was also closely involved with Caledonian Coal Company. Caledonian was a Scottish company with major interests in the northern district mines, its portfolio including the Aberdare, Aberdare Extended, West Wallsend, West Wallsend Killingworth and Waratah collieries. By 1912, when Caledonian was being restructured into an Australian company, Howard Smith had become the dominant shareholder. Howard Smith had also been Caledonian's local agent for many years. McIlwraith McEacharn and Adelaide Steamship also had strong links to collieries in the NSW southern district.[176]

J&A Brown used its own ships to transport its coal to Sydney and to other markets. In 1905 AACo proposed to John Brown, the company chief, that the coal companies should form an association to limit the shipping companies' power. AACo also wanted to see an increase in the price of coal. "Brown had similar ideas, but at first wanted ship–owners excluded until the association was powerful enough to influence them."[177] The AACo proposal went nowhere, but in January 1906 the seven major coal companies in the north (not including Newcastle Wallsend) decided to form an association and met with the shipping companies to discuss selling arrangements. At the meeting in January it was proposed that the coal companies would set a price for all the coal required by the shipping companies' contracts. The shipping companies would set a maximum freight rate for the interstate trade "to keep the consumer price within bounds in order not to encourage outside competition, and were to deal exclusively with the associated collieries for the supply of coal to Victoria, South Australia, Western Australia and Queensland markets.."[178]

The agreement was finalised in September 1906 but was not signed as the vend members had been advised that any signed agreement would expose the companies to the Australian Industries Preservation Act.[179] Associated Northern Collieries (ANC) was the vehicle for the arrangement. The shipping companies agreed to accept the price set by the ANC members and to refrain from buying coal for these markets from any other source. The agreement also involved an increase in the price of coal for 1907 and in the hewing rate paid to the miners. The refusal of Newcastle Wallsend, one of the major producers, to join the cartel could have been a problem for the other producers. However the combination of those other major colliery companies and the shipping companies meant that these companies now controlled a large part of Australian coal production and, even more important, they had a stranglehold on the interstate coal trade.

The vend was well received in the northern district, with the Newcastle Morning Herald acknowledging that while it was not well received by some, it was generally seen as benefiting the district.[180] The newspaper said that the producers did not see the objective of the vend as a way to make coal consumers pay exorbitant prices, but rather to enable them to earn a reasonable profit on the substantial capital that had been invested in their mines and to prevent cut throat competition. "Their (the producers') argument is that immense capital has been sunk in the collieries of the Newcastle and Maitland districts and millions of tons have been drawn from these collieries with little resultant profit." The Herald did warn however that the vend may not prove to be sustainable, with history pointing to the potential for the fight for markets leading to a breakdown in the arrangement. The mine workers were also seen as early beneficiaries, with a rise in the hewing rate tied to the higher coal price.

The Australian Industries Preservation Act was passed by the Federal Government in 1906. Modelled on the US Sherman Act, it was designed to protect Australian industries from unfair competition, prohibiting combinations and monopolies relating to trade and commerce and combinations in restraint of trade.   Federal Parliament debated the Australian Industries Preservation Bill in July 1906. J C (Chris) Watson, the Labor Opposition Leader (who had briefly been Prime Minister in 1904), sang the praises of the vend during that debate, declaring that the agreement between the northern producers was a laudable one. Watson cautioned that "we should take care that nothing is done to prevent such legitimate combination amongst coal mine-owners … to enable them to get a fair profit on their capital invested and to insure fair wages to their employees." Watson said that the industry had been for years "pouring out our natural wealth from the coal-mining districts, to outsiders in particular, at a ruinously low price….at a price which yielded no profit to the owners, to the miners, or to the State as a whole." He believed that "Any system which would overcome that, and operate in the direction of fair and reasonable trading might be welcomed by any person."[181] Watson however changed his opinion in 1907 once he realised that the vend was not just a convenient arrangement between the coal companies to set coal prices which also had benefits for the miners. He now saw the vend as an abuse of market power and encouraged the Government to use the Australian Industries Preservation Act to challenge the arrangement.[182]

Billy Hughes was to play a major part in prosecuting the members of the vend. In 1909 Hughes had been in favour of the vend and spoke in glowing terms of its benefits, saying that it was "…not only a labour saving device, and the best and latest at that, but it marks a notable epoch in the history of production and civilisation. The central ideas of the combine are co-operation and systematisation. It substitutes order for chaos and combination for competition. It takes cognisance of factors utterly ignored by the old barbaric ways of cut-throat competition."[183] However, in 1910, as Commonwealth Attorney General, Hughes launched a prosecution of the members of the vend in the High Court. Justice Isaacs, who had been involved in developing the Industries Preservation Act as Attorney General, heard the case.

Isaacs came down in favour of the Government in 1911, ruling that the vend was in restraint of trade. He was scathing in his assessment of its impact: "I would entertain no doubt whatever the public have borne, are bearing, and will, unless the combination is restrained, continue to bear a heavy detriment in regard to the cost of coal attributable entirely to the existence of the defendants' combination."[184] However on appeal, three judges of the High Court overturned Isaacs' decision, ruling that the prosecution had not established that the arrangement was against the public interest. Referring to the selling price of coal by various shipping companies in various states the Chief Justice said: "In this case, therefore, as well as that of the f.o.b. prices, we find no foundation in the evidence for the conclusion that the prices asked and received by the coal merchants were unreasonable in any sense of the word."[185]

The case was taken on appeal to the Privy Council in London by the Commonwealth in 1913. The Privy Council dismissed the Commonwealth's appeal. The Privy Council was clear in its dismissal of the Commonwealth's appeal: "In their Lordships' opinion the decision appealed against was right, first, because so far as the Crown relied upon sec. 4 (1) (a) and sec. 7 of the Act, there was no evidence (at any rate no satisfactory evidence) of any sinister intention on the part of either colliery proprietors or shipping companies; and secondly, because so far as the Crown relied on sec. 10 there was no evidence (at any rate no sufficient evidence) of injury to the public."[186]

The Campbell Royal Commission of 1919-1920 found that the northern vend had "apparently dissolved" after the decision by Justice Isaacs, and before the High Court appeal decision was brought down.[187] Campbell's report said that there were no further selling prices declared by the northern producers up to January 1916, but that through a tacit understanding the selling price of coals from the Newcastle and Maitland districts was maintained at the level last fixed by the vend until the beginning of 1916. Late in 1915 a threat of strikes was hanging over the heads of the northern producers, with the miners demanding that the producers increase prices so as to increase the miners' hewing rate. The producers did agree to increase prices, and under the sliding scale, wages also increased.

According to John Brown when he addressed his company's annual miners' picnic day in March 1915, the northern producer association (the Colliery Proprietors' Defence Association) was practically wiped out, and he said that prices were being cut by producers.[188] The official Department of Mines statistics however show only a modest drop in the average price for 1915. Average prices received by mines in the north had fallen between 1912 and 1915, but only by around 5%. But the price increase announced by the producers saw the average jump by almost 18% in 1916. However 1915 was also a time when the industry was feeling the effects of the opening of the Panama Canal and the consequent loss of export markets. Production in the north fell that year by around 11%, and would fall by over 15% in 1916 as the impact of the Government's export restrictions and the war hit the industry hard.

Nationalisation and state-owned mines

The Australian Labor Party was born out of the formation of the Labor Electoral Leagues in NSW in 1891. In 1895 the Conference of the NSW Party adopted a Platform which contained the call for government ownership of key elements of the economy: "Ownership by the state or local government bodies of such works as railways, tramways, water supply, public lighting or other works for the good of the community". It was not long before the Conference, by an overwhelming majority, extended this element of the Platform to a range of other industries. The 1896 Platform now called for the nationalisation of all coal, silver, copper and iron mines, and for all iron ore for state use to be produced by state-owned mines. The 1896 Platform also called for the encouragement of agriculture by the establishment of state-owned mills and depots for sugar, grain and produce; and the absorption of the unemployed by the establishment of state farms and labour colonies. 1891 also marked the birth of the Labor Party in Queensland when striking workers met in Barcaldine; the first Labor candidates were elected to the Queensland Parliament in 1893.

The role of government in key industries, including coal mining, was now formally on the public policy agenda. The Trades Union Congress (later the ACTU) passed a resolution in 1902 supporting the nationalisation of key industries, including coal. In 1905 the annual conference of the NSW Labor Party confirmed the desirability of nationalising coal mines, and in July of that year, in preparation for the election, the Federal Party adopted a new Platform which included the nationalisation of monopolies (but which did not call for nationalising coal mining).

Labor came to power in NSW in 1910 with the election of the

McGowan Government, the first NSW Labor Government which had a majority in its own right. One of the new Government's key policies was the development of a State-owned coal mine and steel works. The coal mine would be developed to supply the Government's own needs (primarily for use by the railways), and the steelworks would, according to McGowan, have the potential to satisfy NSW demand, but also demand in other States. The development of a State coal mine was a major new policy for NSW, but fell well short of the nationalisation of coal mining that unions and the Party had been demanding for many years. With the creation of the new vend by the northern coal producers and the shipping companies in 1906 and the subsequent improvement in coal prices and wages, the pressure for nationalisation eased. But following the disastrous 1909 strike in NSW involving union leader Peter Bowling, the dependence of the Government railways on privately owned mines which were subject to potentially lengthy industrial disputes was highlighted.

Soon after its election, the NSW Government under McGowan proceeded to search for suitable locations for its new coal mine. The area around Muswellbrook was the early target by the Department of Mines, but the focus changed to the Lithgow area and in 1911 the Minister for Mines announced that a suitable site had been identified near Lithgow, with good access to the rail system.[189]

In 1912 the NSW Government passed the State Coal Mines Act which authorised the Government to establish or purchase coal mines for the purpose of supplying the Railways and other users, and to secure Crown land or private land. Finally in 1916, the go-ahead was given and development of the new mine commenced in the vicinity of other operating mines in Lithgow. However after progress had been made on the sinking of a shaft, financial constraints forced the NSW Government to suspend further work in 1917. In 1920 legislation was passed to pass control of the mine to the Chief Commissioner of Railways, and development of the mine proceeded in 1921, with production of coal commencing late that year. The mine was employing 118 men in 1921, and this jumped to 337 in 1922, making the State Coal Mine by far the biggest mining workforce in the western district. With employment growing to 501 in 1923, the mine was now rivalling some of the large mines in the north.[190]

State ownership and operation of coal mines would be significantly expanded in NSW after World War Two with the expansion of the mines controlled by the State Coal Mines Control Authority to supply the NSW railways and the new power stations developed by the NSW Electricity Commission.

Queensland Labor's agenda starts in earnest

The Labor Party gained power in its own right in Queensland in 1915, having campaigned on a radical program including nationalisation of key industries. The election of the Labor Government under Thomas Ryan that year saw major initiatives implemented in a short space of time. In fact the changes were such that by 1917 Ryan had become "the socialist ogre so far as Australia's daily conservative newspapers and local and foreign capitalists were concerned."[191] Ryan tried to abolish the Queensland Legislative Council in 1917, but was unsuccessful, although this was achieved by the Labor Government in 1921 led by E G (Ted) Theodore, who had served as Treasurer in the Ryan Government. Ryan and Theodore were able to implement major elements of their program in the period between 1915 and 1923. A board to decide the price of sugar was established, a workers' compensation scheme run by the Government was commenced, and a number of state-owned mines and other business established or taken over. Ryan also had plans for a state-owned steel works and shipping line, but these never proceeded.

The new Government's first major foray into mining was the takeover of the troubled Chillagoe metal mines and smelter (located just over 200 kilometres west of Cairns) from the Chillagoe Company in 1915, and at the end of that year the Government announced that it had negotiated to purchase the Warra coal mine, near Dalby. There were also reports in early 1916 that the government was negotiating to buy coal mines in the West Moreton district but these deals did not eventuate.[192]

The Warra mine was the first operating state-owned coal mine and it was not the end of the Government's interest in the coal mining business.[193] The Government carried out drilling in several areas to assess their potential for mining and within the next few years it was developing mines at Styx River between Rockhampton and Mackay, Baralaba (about 140 km west of Rockhampton) and the Bowen State Mine at Collinsville at the northern end of what we now know as the Bowen Basin. And in 1923 it added to its portfolio with the purchase of the Mt Mulligan coal mine on the Atherton Tablelands west of Cairns.

The Warra mine proved to be too difficult and costly and after the Government was re-elected in 1918, it abandoned the project. The Bowen and Mt Mulligan mines went on to have a long life, but Baralaba was closed in 1928 following major floods in the region which fatally damaged the mine, and Styx was closed in 1934 following an explosion and fire. The Bowen State Mine was owned and run by the Government until it was sold in1961, and mining in the area continues today at Collinsville, one of the Glencore stable of mines. Mt Mulligan operated until its closure in 1957, having been the scene of one of the worst mining disasters in Australia in 1921 when an explosion in the mine killed 75 people.

An iron and steel works was also planned by the Labor Government which appointed a Royal Commission in 1917 to assess its potential. The Commission delivered an interim report in 1917 and a final report in 1918, these reports being optimistic about the potential for a new industry using local coal and potentially local iron ore. A site was selected south of Bowen which would use coal from the new Bowen State mine being developed about 100 kilometres north. However the problems were too great and the project had been abandoned by 1920, leaving some coke ovens as the only ongoing legacy.

Minmi – an early enterprise bargaining agreement

The problem of intermittency – the irregular availability of work – on our coalfields existed for much of the life of the industry up until World War Two. Mine workers would be lucky to work a full week or fortnight, with perhaps 4-6 days per fortnight common. Most collieries worked far fewer days per year than the maximum allowed under industrial awards and mining regulations. We will return to the issue of intermittency in chapter 3.

At one major mine in the northern district, the "Coal Baron", John Brown, head of J&A Brown, struck an agreement in 1913 with the workers at his Minmi operations west of Newcastle, one of the major aims of which was to provide regular work to the mine workers. The so-called Minmi Agreement was for 5 years and gave mine workers an increase in pay, a dispute arbitration process, a commitment from the mine workers that they would not strike while any dispute was being arbitrated, and a commitment from the company that it would do all it could to ensure continuous work for the mine workers. As the agreement was negotiated between the Minmi Lodge of the Miners' Federation and the company without the involvement of the district union, the Lodge was promptly expelled from the union. If the Hunter River District Colliery Proprietors' Defence Association were to be disbanded, and in the event of price competition between the companies leading to price cuts, the agreement guaranteed the pay levels until prices reached a certain level, at which time there would be a renegotiation of the pay rates.[194] The Association was the body representing the northern producers on industrial matters, and had seen the withdrawal of some major producers including Caledonian, Abermain and New Lambton.

One year later reports of the success of the agreement indicated that both the company and the employees were happy, with little or no disputation occurring at the Minmi collieries. That was not the case at Brown's other collieries, for example with the Pelaw Main colliery, at which Brown said that there had been 13-15 disputes in the previous three months alone.[195] But everything for Minmi would not continue to be so plain sailing. After hitting around 127,000 tons in 1914, production from Minmi would sink to only 19,000 tons in 1917, before recovering to 75,000 tons in 1918. J&A Brown continued to operate pits at Minmi until 3 January 1924, when the miners at the Back Creek pit took the day off work when one of their long serving colleagues died on the job that morning.

As this was the first day back at work following the Christmas break, John Brown ordered the men back to work, threatening to close the mine if they did not. "The miners buried their mate. John Brown buried the town."[196] Back Creek was the last pit worked at Minmi, a town that at its peak had a population of 6000 to 7000. The Duckenfield pit had closed in September 1923 following industrial disputes and now Back Creek was also closed. By the time of their closure, the pits around Minmi had been mined for many years and had seen the best coal extracted. As Christopher Jay wrote in The Coal Masters: "If you wanted to make an example of someone, it would make more business sense to do it with the less economic properties. The Duckenfield and Back Creek miners were to find this out."[197] J&A Brown had also developed a new mine west of Minmi called Stockrington, with production commencing in 1921. But following a strike, John Brown closed the mine and it would remain shut until after his death.

The Commonwealth becomes a coal industry regulator

World War One was a watershed for the coal industry in a number of ways, but principally because it marked the entry of the Commonwealth government into the direct regulation and operation of the industry. The early war years also saw another major development - the formation of the Australasian Coal and Shale Employees' Association, later to become the Australasian Coal and Shale Employees' Federation or Miners' Federation as it was commonly known. The Association was launched in March 1915, the month before the Gallipoli landing by Allied forces and the military campaign which is often seen as the event which marks Australia's "coming of age" as a nation. The Association brought together the miners unions in NSW, Queensland and Tasmania, and aimed to extend its coverage to other parts of Australia and to New Zealand.

The War Precautions Act passed in 1915 allowed the Commonwealth Government to restrict the export of key commodities including coal. Coal exports from NSW, the only significant exporting state, declined in the next few years, dropping from 2.6 million tons in 1914 to only 725,000 tons in 1918, before rising again after the war ended. As we explore later in this book, following World War Two the Commonwealth and NSW governments passed almost identical legislation in 1947 to establish the Joint Coal Board and the Coal Industry Tribunal. The CIT's role was to extend to the Queensland industry and the tribunal continued to exist as a separate industrial relations tribunal outside the jurisdiction of the mainstream tribunal until 1995, much to the displeasure of the coal industry employers, who, from the 1980s, fought to abolish it and bring coal industrial relations under the then Australian Industrial Relations Commission.

This Commonwealth involvement in coal industrial relations can be traced back to a major strike in 1916 which led Prime Minister Billy Hughes to appoint Justice Edmunds as a special commissioner under the War Precautions Act to arbitrate on the miners' claims. Hughes first sought to appoint Justice Higgins of the Arbitration Court, but Higgins declined, later stating that a condition of his appointment by Hughes would have been to agree to the miners' claims.[198]

After a quick one day hearing in December 1916, Edmunds granted the miners' demands including a working day of 8 hours bank to bank (8 hours from the time the first miner arrived at the pit top to the time the last miner arrived back at the pit top), an increase in wages of 20% and the abolition of the sliding scale principle on which wages had been based for decades. This decision saw the most dramatic improvement in miners' wages and conditions since the 1880s. It also resulted in a major lift in the authority of the Federation's leaders, James Baddeley and Albert Willis, and for the next fifteen years was used to demonstrate that direct industrial action rather than arbitration was needed to gain real results for miners.[199]

The Edmunds decision also included a provision that there should be no further industrial action for the following three years or longer if the War continued beyond that time. Edmunds also ruled that no action was to be taken by any party that would destroy the peace of the industry. As has often occurred in the industry throughout its history, it was not long before that sort of "no further action' ruling was breached. However from that point in the early twentieth century, the Commonwealth, for better or worse, had become a major player in the Australian coal industry.

Sir Colin Davidson, who conducted the 1945-1946 inquiry into the industry, and also headed the Royal Commission into the NSW coal industry in 1929, was scathing of the way in which Hughes handled this dispute in 1916. Writing in 1953, he noted that Hughes' action had taken the matter out of the hands of the Federal Arbitration Court which had made an interim award, but which was disobeyed by the Miners' Federation. 'Thereby a precedent was created which has led to one demand after another, not only for similar political intervention, but also for removal of Judges or judicial officers who have failed to grant in full various claims of the union. Probably no more insidious blow could have been struck at the whole concept of arbitration and the efficacy of its tribunals.'[200]

The intervention of the Commonwealth in 1916 was a major change for the industry, and marked the commencement of a new era in which the Commonwealth would play a major role in coal mining, initially in terms of industrial relations, and later in the restructuring of the industry in NSW in the post-World War Two years, and subsequently in areas including foreign investment and coal pricing and marketing. There was also another intervention into the industry in 1917, this time by the NSW Government.

The northern coal lockout of 1929-1930 and the 1949 national coal strike live on as perhaps the two most famous or infamous disputes in the coal industry. But 1917 also saw another historic strike, known at the time as the Big Strike. The 1917 strike began in August in the NSW Railways and quickly spread to coal mining, the waterside and industries related to the railways through sympathy strikes and unions declaring certain products "black". Some strikes were also protests against the employment of non-union workers. The Miners' Federation refused to handle coal supplied to the railways, and this action brought the coal industry into the strike. By the end of August there were thousands of workers, "farmers and their sons", who were living in tents at the Sydney Cricket Ground and at Taronga Zoo and who had been recruited by the NSW and Commonwealth governments to work in the government trams and trains and on the wharves.[201]

The NSW Government said that the strike was a conspiracy to overthrow the government and to undermine the war effort. And the government's position in relation to the coal industry and the union was made crystal clear, with the Acting Premier saying in September: "When the Coal Miners' Federation deliberately challenges the community and states that it will not cut coal to meet our necessities unless its own terms in detail are granted, we can only accept this challenge in the same way as we accepted the challenge of the railway employees. Just as we have proved to the community that we can run the railways without the concurrence of the strike committee, so we must establish that we can get sufficient coal for the needs of the community without the consent of the Coal Miners' Federation."[202] The government commandeered all coal stocks in the northern district and then took over all coal mines and began operating them with non-union labour. To ease the way for non-union workers the government also amended the Coal Mines Regulation Act to allow men with little experience to work at the coal face underground.

The railway strike ended in September, and the coal mine workers returned to work in October. As part of the settlement of the coal strike, the Government stipulated that the Federation's members had to work with non-union workers who had been recruited during the strike and to admit them to the union if they wished to join. The strike and its settlement created a great deal of bitterness in the industry, but over time the non-union workers tended to drift out of the industry.

Some previous major disputes had seen the miners' union effectively crushed and rendered irrelevant for some years afterwards. However, while the 1917 strike was a failure, the Federation was able to continue as an effective and powerful union.[203] The NSW Government's actions in taking over the coal mines and operating some of them with non-union workers was certainly a first for the industry, but it could perhaps be seen as setting a precedent for the action of the Commonwealth Government in 1949 when the Army would be sent in the operate mines in NSW.

Birth of Australia's steel industry

The history of the Australian iron and steel industry goes back to 1848 when the first iron works was established in Mittagong in the NSW southern highlands. An iron works commenced operating in Lithgow in 1875, and the first steel production in Australia was in Lithgow in 1900. But the birth of the modern Australian steel industry was in 1915 with the opening of the BHP steelworks in Newcastle.

In line with the Labor movement's desire to see key industries owned and operated by government in the era before World War One, in 1912 the NSW Government appointed three of its Ministers to look into the establishment of a state-owned iron and steel works. This led to the Newcastle Iron and Steel Works Bill being introduced into the Legislative Assembly where it was passed by a majority of members, but the Bill was rejected in the Legislative Council. However in May of 1912, BHP's chief, Guillaume Delprat, approached the NSW Treasurer, J H Cann, with a request that the NSW Government assist BHP to set up in Newcastle. Delprat's letter also mentioned the option of establishing a steelworks in South Australia, where the company had its Iron Knob iron ore deposit, a not-so-subtle reminder that the company was not necessarily locked into Newcastle. Iron Knob was already supplying ore used as a flux to BHP's silver-lead smelter in Port Pirie.

With Newcastle's coal industry having lost many jobs over the past decade as the coal industry saw a surge of development in the South Maitland coalfield to tap the Greta seams, the prospect of a major boost to jobs in Newcastle was something the Government could not ignore, even if they were not jobs in a state-owned steelworks. With no prospect of the Bill passing the Upper House, and doubts about the ability of the Government to fund such a major facility, Premier McGowan advised BHP that "any industry that may be established in the State of New South Wales will receive encouragement and consideration from my Government." This not only ended the push for a nationalised steel industry in NSW, but also opened the door to the Big Australian, as it later came to be known, to expand beyond its limited operations in Broken Hill and become one of the powerhouses of Australian mining and industrial development.

The Newcastle Iron and Steel Works Act which gave the necessary approvals for the establishment of the Newcastle steelworks was passed in 1912, despite being opposed by a number of Labor members of Parliament. The following year the Conference of the NSW Labor Party passed a motion censuring the Government for providing assistance to BHP to establish the new plant. The Newcastle steelworks commenced production in 1915 and became a new and significant market for NSW coal producers. During the 1920s, BHP went on to enter the coal mining industry, and by the mid-1930s had become one of the major producers in NSW.

Stockton Colliery Newcastle 1887. (NSW State Archives)

Greta seam - Stanford Merthyr Colliery early 1900s. (Courtesy Mine Super)

3

The Lost Years: between the wars

In contrast to the situation during World War Two, when the coal industry would prove unable to meet the challenge of higher demand for coal for the war effort, World War One saw the industry contract as demand for coal declined. One factor in this contraction was the loss of export markets, a direct result of the Commonwealth Government's restrictions on coal exports; another factor was the opening of the Panama Canal in August 1914, only days after Britain declared war on Germany, causing our industry to lose markets to other exporters.

Production in NSW in the pre-war years peaked at over 10 million tons in 1913. During the years 1916 and 1917 production was down by almost 20% to only 8.1 and 8.3 million tons respectively. It rose to just over 9 million tons in 1918, before easing back to 8.6 million tons in 1919, and then rose again to 10.7 million tons in 1920. NSW exports fell from 2 million tons in 1915 to only 725,000 tons in 1918, before recovering to 1 million tons in 1919 and 2.7 million tons in 1920. Employment in NSW mines peaked in 1914 at just over 20,000. By 1918 numbers had fallen by around 3,000. In Queensland production averaged just over 1 million tons a year from 1913 to 1915, and then fell to just over 900,000 tons in 1916, before recovering again to over 1 million tons in 1917. Employment in 1915 was just over 2500, but by 1916 had fallen by almost 500. Numbers recovered somewhat in 1917 and 1918 to over 2200. The reduction in employment in both states reflected the lower production in the industry and also the fact that many miners had enlisted into the Australian military forces. Using their skills learnt in Australia, some of the miners were a key part of the dangerous tunnelling units at Gallipoli and on the Western Front in France and Belgium.

The 1919 Royal Commission

The Miners' Federation, recognising the new and powerful role which the Commonwealth had assumed in the regulation of the industry since 1916, had been pushing the Federal Government to hold a Royal Commission into the NSW coal industry, but without success. In April 1919, the NSW Labor Government led by Premier Holman, stunned the industry with its own decision to appoint a Royal Commission to be chaired by James Campbell K.C. Campbell was a barrister who went on to be appointed to the NSW Supreme Court in 1922 and to conduct a number of other Royal Commissions and inquiries during the 1920s for the NSW, Federal and Queensland Governments. Charles McDonald, the secretary of the Northern Colliery Proprietors' Association, said that the employers had recently attended a conference in Melbourne convened by the Federal Government where it had been provisionally agreed to increase wages. Expressing surprise at the State Government's action, McDonald said that "The Federal Government have dealt with the coal industry since war was declared in 1914, and we assume they will still deal with it."[204] James Baddeley, president of the Miners' Federation, said that all of the union's negotiations had been through the Federal Government, and that as far as they were concerned, "we are going right ahead with them. It is no concern of ours what the State Government have done or are doing."[205]

The NSW Government ignored the industry's concerns and gave Campbell wide-ranging terms of reference. Some of the key issues that the Commission was to examine included who were the companies involved in the industry, what sort of profits they were making, and what changes in the industry, if any, were desirable. Campbell's first report was released in 1919, addressing ten of its terms of reference; his second report was finalised in 1920. Campbell's findings and conclusions appear at different points throughout these reports, making it difficult to summarise his overall assessments of the industry. He did however conclude that there was "no satisfactory reason" to make any fundamental changes to the industry in terms of its system of ownership of collieries. But in terms of the relations between employers and employees, and in the way in which collieries were managed, he saw a serious need for improvement.[206] One of his major conclusions was that much of the trouble affecting the industry would disappear if adequate dispute resolution mechanisms were established.[207] Whether the industry needed a Royal Commission to make this earthshattering conclusion is debatable.

One of the key features of the 1919 report is a summary of the value of capital and profits of colliery companies for the years from 1914 to 1918. Campbell did not attempt to do any detailed analysis of profitability, but at least was able to demonstrate to the Government that he had met that part of the terms of reference. Profitability varied widely between companies and over time, ranging from losses in some years for a few collieries to very strong profits for others. The figures in the report do not lend themselves to charges that the industry as a whole was making exorbitant profits. For example, for the ten largest collieries in the northern district, the average rate of return on capital was around 11% in 1914 (this figure was for only seven of the ten), rising slightly to around 12% in 1915 (not including one company which made a loss), and falling to around 9% in 1916. However these collieries overall saw better times in 1917 and 1918 with a return of close to 17%. In 1918 the returns ranged from around 3% for two companies to 37% for one company (but that company's results included substantial profits from non-coal operations).

Campbell also made some interesting findings about the selling cartels or vend arrangements in the period after the famous northern Vend case. He concluded that the Vend had apparently been dissolved without the producers waiting for the High Court decision. So while Campbell concluded that the vend ceased to exist in a formal sense, he found that the selling price of coal from the Newcastle and Maitland areas had been maintained through a tacit understanding at the same level as had been fixed by the Vend prior to the High Court judgement [208].

Under the threat of strike action from the miners, the northern district employers had moved to lift the selling price late in 1915, and under the sliding scale for wages, miners' wages were increased. This was to be the first step in some significant increases in wages and prices over the following few years. Campbell found that the vend arrangements in the southern and western districts in NSW continued to operate, but he was quite restrained in his assessment of their impact. In fact he concluded that there was no evidence that the vends operated by the Southern Coal Owners' Association and the Lithgow Coal Association had any adverse effects on the public or on coal consumers. He did warn however that the Government would be wise to keep them under close watch as collective arrangements like these carry the potential to cause harm.[209]

The other key aspect of selling arrangements was the link between the coal producing companies and the shipping companies. Campbell said that after the northern vend was formally abandoned, the shipping companies "continued in a general way their relations with the Maitland collieries" for which they had acted as selling agents.[210] His report gave some details of the relationships between the major shipping companies and some of the major coal producers. The major companies concerned were Caledonian Colleries, Hebburn Colliery, Abermain and Seaham Collieries, and Hetton Collieries. Caledonian Collieries Limited had been formed in 1913 out of the old Caledonian Coal Company Limited, a company registered in Scotland and which had Howard Smith as its local agent. Campbell concluded that Howard Smith was entitled to receive all of the share capital in the new company, bringing Caledonian under full control.

Hebburn Colliery Company was formed in 1914 when AACo sold the Hebburn colliery and Abermain railway assets to the new company.[211] AACo continued its interests in the industry through a major shareholding in Hebburn, but the dominant shareholder in the new company was the shipping company Huddart Parker.[212] The other key links were between Adelaide Steamship Company and the Abermain and Seaham collieries, and between McIlwraith McEacharn and Hetton Collieries which operated the Bellbird colliery. J&A Brown continued to operate its own shipping. These links were to continue, in some cases for many years, but did become less powerful as new companies entered the industry, including BHP in the 1920s and 1930s, and as the Joint Coal Board's mining operations developed following World War Two.

Commonwealth passes new industrial legislation

The Commonwealth Government's intervention in the coal industry through the appointment of Justice Edmunds under wartime legislation marked the entry of the national government into the detailed regulation of the coal industry. In another wartime move, the NSW Government intervened, in 1918 when the Acting Premier issued orders under the War Precautions Act "acquiring all New South Wales coal for the Commonwealth" and raising wage rates.[213]

In 1920, the Industrial Peace Acts were passed providing for industrial tribunals to be established to regulate conditions for various industries. Four tribunals were set up for the mining industry under these Acts: one for the miners, one for the engine drivers and firemen, one for engineers, and one for the coke industry. Justice Charles Hibble, who had been involved in chairing voluntary tribunals for the coal industry in the pre-war years, was appointed to chair these mining tribunals. Hibble was involved in a number of major industrial cases in the 1920s. In one major case in 1923 he arbitrated on an application by the employers in NSW, Queensland, Victoria and Tasmania to reduce the wages of mine workers by one third, largely on the basis that the cost of living had fallen. In rejecting the employers' application, Hibble referred to the intermittency in the industry which was a factor when the wage rates were previously set in 1920, and gave this as one of the reasons for his decision.[214]   At the end of his decision, Hibble rounded on the industry in similar terms to those that would appear in the future, for example in the Davidson Royal Commission in 1929-30 and in Davidson's Coal Industry Inquiry report in 1946. Hibble said that while he knew "only too well" that it was "quite futile to expect a huge industry like the coal industry, with all its varying conditions and methods, to proceed without an occasional stoppage at one pit or the other, … a position has developed, and especially in the main coal centre, namely the Northern field... which has become intolerable." He said he was referring to "the persistent and ever-recurring stoppages and dargs at this or that mine over grievances which as a rule are of a most flimsy character, and which in no case call for the holding up of an industry. The absurd futility of this course ought to be apparent, and if means can possibly be adopted it should be ended. The miners officially have adopted the arbitral method of dispute settlement, although in actual practice it is very often strike first and arbitrate afterwards. I am convinced that the great bulk of the employees at the mines are opposed to these sectional stoppages, but it is the case of the many led by the few. Disciplinary authority rests with the council of the federation, and the officers of that body have a duty to perform, not only to their own members, but to the general community, by absolutely insisting that as far as possible, these stoppages and interferences with the industry must cease." Hibble went as far as warning that the award system was at risk: "The time has assuredly arrived when it will have to be seriously considered how much longer it can be permitted to allow the benefits and protection of awards to run concurrently with methods completely at variance with the principles and understanding upon which those awards were made."

The Hibble Coal Tribunal operated throughout the 1920s, but by the end of the decade its future was looking uncertain to say the least. Hibble fell ill in mid-1929, although did get involved in an unsuccessful attempt to broker an agreement between the employers and the unions during the Northern lockout. He resigned in 1930 as head of the Coal Tribunal because of his health; he was not replaced and the Commonwealth Arbitration Court stepped to become the coal industry's industrial umpire.

Bellbird Disaster and the 1925-26 Royal Commission

Just before midday on Saturday September 1, 1923 (or just before 2pm on the east coast of Australia), the Japanese port of Yokohama and surrounding areas were devastated by an earthquake and tsunami which resulted in a death toll of around 140,000 people. In what is known as the Great Kanto Earthquake, many of the people were killed as a result of fires which raged after the earthquake. While those fires were raging, thousands of kilometres away in Australia, a tragedy was unfolding at Bellbird near Cessnock in the NSW Hunter Valley. The Bellbird mine was the scene of the first major NSW coal industry tragedy of the 1920s.

At around 1.30pm local time, a fire was detected in the mine; the fire led to a series of explosions that day and also early the next day, with 21 miners killed by carbon monoxide poisoning. Six bodies were not initially recovered from the mine, although five bodies were later recovered when the mine re-opened. Those killed included John Brown, the manager of the Aberdare Colliery, who had been participating in the rescue efforts. The Bellbird explosions, around 7 in all, took place from around 2pm on September 1 to 3am on September 2. The carbon monoxide was found to have been generated by the fires which in turn were caused by the ignition of coal gas from overheated heaps of coal within the mine.

The Bellbird disaster was a massive blow for the village of Bellbird and the Cessnock district. On Monday September 3, an estimated 25,000 people lined the route of the funeral procession for the dead miners in Cessnock, with hundreds of miners part of the funeral procession. But the tragedy did lead to some critical steps forward for the coal industry in terms of mine safety and rescue. Queensland had led the way with the first coal mines rescue station set up in north Ipswich; the station was completed and equipped in 1912. The north Ipswich station was the initiative of the Mines Department in November 1909, but by 1919 it was believed that a new and better equipped station should be developed. The new station in Booval opened in 1923, located on Crown land; its running costs were shared by the Mines Department, the State Government Insurance Office and the Queensland Coal Owners' Association.[215]

The Bellbird tragedy led to the passage of legislation in NSW in 1925 which provided for the establishment of a mines rescue system in NSW coal and shale mines. Rescue stations were now to be established in four coal mining districts (at Abermain in the South Maitland district, at Cockle Creek in the Newcastle district, at Lithgow in the west and at Bellambi in the south), with these stations equipped with a range of specialised equipment, breathing apparatus and vehicles. Mines rescue brigades were also to be established at certain mines where there was no permanent rescue corps in the district.

Bellbird also was an important factor in the appointment of a NSW

Royal Commission in 1925 to investigate safety in the industry. Under the chairmanship of Justice Edmunds, the Commission reported in 1926, but recommended the continued use of naked lights in collieries where no problems had been experienced in the previous year, or where there was no likelihood of any problems. It pointed to the continued use of naked lights in Britain and the USA and said that "Great weight of authority throughout the world is in favour of retention…upon some defined principles."[216] The Commission noted the superior light given off by naked lights, their convenience, and also the fears expressed by miners of "miners' mystagmus" (an involuntary eye movement) which miners claimed was caused by the use of safety lights. The naked light saga again dragged on into the 1930s.

The Commission did however lead to some important changes to the Coal Mines Regulation Act in 1926, including provisions for improved ventilation in mines, the prohibition of furnaces in mines (although this did not apply to small mines), and provisions for stone dusting in roadways which carried electrical cables and for the removal or treatment of coal dust in roadways. Stone dusting and the management of coal dust were critical moves forward in minimising the threat of explosions underground. The 1926 Act also required producers to provide sanitary facilities above and below ground and to build bath and change houses. However as we will see from the Joint Coal Board surveys after World War Two, it would take many years for major improvements in sanitary facilities and bath houses to be implemented.

Queensland State mines – one failure, and some progress

Following its election in 1915, the Labor Government in Queensland would have been pleased to see that some of its plans for development of State-owned and operated coal mines were proceeding by the end of the War. By the end of 1918, the development of the Bowen State Mine was underway, and testing and drilling were underway or planned at Styx River and Baralaba. However while the Warra mine located about 50 kilometres north west of Dalby on the Darling Downs was operating, it was soon to be shut down.

Problems with the original Warra mine had forced the Government to develop a new mine in 1916, but problems were again encountered with water inflows and boiler failures.[217] By 1917, with two years having elapsed since the new Labor Government had been elected on a platform which included state coal mines to produce better quality and cheaper coal for the railways, progress with Warra had been slow. In the election year of 1918 "the Government was desperate for some signs of success" and the mine manager "was therefore instructed to commence production immediately from the thin seam already reached, hoisting coal up the air shaft so that sinking could continue in the main shaft... Almost 4,000 tonnes was extracted by this inconvenient route and burned by the Railway Department or in the pit boilers, thus stifling public criticism at a politically sensitive time."[218]

By 1918 the mine had received delivery of new plant and machinery and was able to install electrical power plant and coal conveying machinery, build a dam for water supply, and complete the construction of offices and a change house for the miners. These investments were planned to allow the mine to double or triple its output. However in 1919 the Government suspended production at Warra and soon after decided to transfer much of the machinery and equipment and the mine manager to the new mine at Bowen. Warra produced only around 11,000 tons of coal over its short life for the railways on the Darling Downs, and it had been a costly exercise for the Government. The 80 Warra mine workers offered to continue to operate the mine on a tribute basis, under which they would pay the Government a royalty for coal produced, or to hire its machinery and equipment and access the coal seam by sinking a shaft on nearby private land. The Government however rejected these proposals. Historian Ray Whitmore concluded that the closure of Warra and the transfer of the machinery to Bowen "marked the end of an ill-conceived, mismanaged State initiative which should never have been started in the first place and was then allowed to continue for far too long before being abandoned."[219]

Before World War One, Mt Mulligan was the major coal mine in north Queensland dedicated to supplying the Chillagoe metal smelters.

There were also privately-owned mines operating in the Bowen coalfield, but the railways in the northern region relied largely on coal shipped from the south through the port of Townsville, and problems in NSW had created shortages of coke for the smelters in the north and central regions of the state. The Bowen coalfield "offered an ideal opportunity for proving that the State could develop and run coal mines efficiently and profitably" and in 1917 the Government geologist, Benjamin Dunstan, was given the task of selecting the site for a state mine.[220] With the competitive threat of a major new Government coal mine, the three private coal mining operations decided to merge their operations into a new company, Bowen Coal Mines Consolidated Limited. The new company and the Government's Bowen State Mine and would dominate the coal mining industry in the Collinsville area for the next forty years.

The development of the Bowen State Mine commenced in 1918, with the Department reporting that a lot of time had been required that year to obtain the necessary machinery and equipment, and that the construction of the railway was expected to be sufficiently advanced to warrant commencing more extensive operations early in 1919. By 1920, the Bowen State Mine had progressed to the stage where further development of the mine was dependent on the completion of the railway line to link the mine to the main rail system. That rail link was completed in 1922, allowing the mine to begin full production. The plan at Bowen was to mine the Garrick seam, which was better suited to making coke, and the Bowen seam, which was suitable for use by the railways.[221]

At the Styx River site, a shaft was sunk in 1918 to test the quality of the coal, and railway construction was well advanced. The Mines Department's annual report for the 1919 year said that the coal quality was exceptional. The same report also stated that coal from the first shaft which had been sunk at Baralaba on the Dawson coalfield had been tested exhaustively, but had proven unsuitable for use by the railways, and could only be used by locomotives if it was blended with more volatile coal. That same year a shaft was commenced at Torbanlea near Maryborough, but the results were not promising enough to warrant developing a colliery on that site, and the focus was moving to another site just to the south.

In 1922 the Government mine at Styx was producing a small quantity of coal from the prospecting shaft, but the shaft was abandoned soon after, with the Government intending to open a colliery a few kilometres north. By 1925 Mines Department annual report gave a positive run down of developments at Styx, which now employed 77 men. A new ventilation fan was installed that year, and two coal cutting machines had also been installed, but with plans to replace the coal cutters with "air driven breast machines of the longwall type"; plans also existed for mechanical haulage to be installed in the mine. The Department said that coal from Styx was excellent steaming coal, and that it was anticipated that in the near future coal for gas-making in the north of the State would be supplied from this mine.

In the early 1920s, the Queensland Government was also planning to develop a steel works and coke works south of Bowen, with plans in the air for facilities in Bowen to enable the export of coal. The steel works failed to proceed, but a small coke works was operating by 1933.[222] At Baralaba, the initial development had been unsuccessful, and a new shaft was sunk nearby and coal provided to the Railways. By 1925, Baralaba was being affected by the closure of a major customer, the Mt Morgan mine and smelter, but managed to produce just over 30,000 tons for the year with an average workforce of 84. Baralaba closed in 1928 following flooding of the mine.

The Mt Mulligan mine, near Mareeba on the Atherton Tablelands and about 100 kilometres west of Cairns, became another state-owned mine when it was taken over by the Government in 1923. Mt Mulligan had been developed from 1911 by the Chillagoe Company to supply its smelting operations about 45 kilometres away. The Government had taken over the assets of the Chillagoe Company in 1919, but left the coal mine in company ownership. The Chillagoe mine and smelters had had a chequered history, closing at times and making significant losses. The smelters had closed in 1914, and with them the Mt Mulligan mine. However Mt Mulligan was back in business in 1915, with the railway now being extended to the mine, giving it access to other markets in north Queensland.

Mt Mulligan tragically became the scene of the worst mining disaster in the State's history in September 1921. An explosion killed 75 men working in the mine, these men accounting for about one in four of the total population of the nearby town. A Royal Commission set up to investigate the disaster concluded that the cause of the explosion was the ignition of an explosive on top of a block of machine cut coal. The Commission did not know whether this ignition was accidental or intended, but the result was an explosion of coal dust. The mine's poor safety standards meant that dust levels in the mine were dangerously high. A litany of failings by the mine was uncovered by the Commission, and its findings led to a major overhaul of Queensland mine safety legislation in 1925, with the passage of the Coal Mining Act; previously, coal and metalliferous mining had been regulated under one Act. Now coal had its own Act which had new provisions including the use of safety lamps, rules on explosives, ventilation and stone dusting, the establishment of mines rescue brigades, and the requirement for mines inspectors to have practical mining experience.

Mt Mulligan was not a profitable investment for the Government and by 1929 losses forced the Government to propose closing the mine. However, following negotiations with the Government, the workers at the mine agreed to take over operation of the mine on what was called a tribute basis, with a payment being made to the Government for each ton of coal produced. By 1930, although Mt Mulligan was still under State ownership, the only State operated coal mines were Bowen and Styx. Mt Mulligan continued to operate until 1957. Styx (then called Ogmore) closed in 1963 and the Government closed the Bowen State Mine and sold it to Dacon Collieries in 1961.

The Northern Lockout

It is generally accepted that the Great Depression commenced with the crash of Wall Street in October 1929, but the northern NSW coal industry and the communities dependent on coal suffered their own depression a little earlier. As NSW Premier Jack Lang observed "One of the fallacies about the Depression is the widely held view that it started with a fall in wool and wheat prices. This is not so. They came later. The first impact occurred in the coal mining industry. It was the critical industry. It registered the first mass misery and suffering seen in this country for a third of a century.'[223] Lang went on to say that with the commencement of the lockout on 2 March 1929 'The Great Lockout had started. The pit whistles no longer sounded at night. For the miners' families there remained only their family endowment and the dole. The Depression had started in New South Wales."[224]

In March 1929 all the major northern district coal companies closed their mines, locking out around 10,000 workers; the lockout was to last for 15 months. The decision of the companies was a harsh and drastic one and devastated the communities involved in the Newcastle and Maitland coalfields. However it was a decision which was the culmination of events over the preceding few years, and in particular the events of 1928. Looking back at the 1920s, it is not hard to feel that there was a distinct sense of unreality within the industry in NSW. The 1920s saw costs in the industry rise, new mines open and production capacity and employment increase. In 1920 there were 138 coal mines in NSW listed as having employees. By 1928 this had increased to 164, with 60 of these employing 10 or fewer workers, and 26 employing between 11 and 50. The corresponding numbers for 1920 were 47 employing 10 or less and 21 employing between 11 and 50, so the bulk of the increase over this period was in small mines, typically requiring only modest amounts of capital investment.

By the start of 1927, the situation in NSW was looking bleak. Exports to other Australian states and to New Zealand were falling. Prices also rose during the 1920s, and by 1927 the industry had begun to see competition from other countries for our local markets. In September 1927 the Northern Coal Association, representing Newcastle producers, even proposed that the Government should impose a duty on coal imports to render imports uncompetitive.[225] Employment in the NSW industry had risen from just over 18000 in 1919 to almost 25000 in 1926. The official figures for 1927 show job numbers falling slightly in 1927, but the big drop was in 1928 when about one in eight miners lost their jobs, a total of around 3000 men. And of course much bigger falls were to come with the northern lockout in March 1929, followed by the impact of the Great Depression in the following years.

In 1928 the Federation held its Second Miners' Convention, with its attitude to the arbitration system one of the major issues. The miners were split between those prepared to work within the system, and those wishing to work outside it by pursuing direct action as the way to secure their claims. A C Willis, the Federation's national secretary until his move into the NSW Upper House in 1925, argued that there was "never a more inopportune time in the history of the organisation in which to have a general strike" and opposed the union abandoning the industrial tribunal which operated at that time.[226] On the other hand, Thomas Hoare, the northern district president, said that there were already 3,000 men in the district "totally unemployed and starving", with thousands of others partially employed. Hoare was in favour of direct action and appealed to the miners, saying "If we do as I suggest, I admit it is only extending the starvation, but instead of one man starving six will starve. If the principle is right, however, then that will be right."[227] Hoare and the direct action supporters won the day, setting the scene for the Federation's stance in the following year.

The coal producers were promoting the need for reductions in coal prices and wages to counteract the industry's falling production. With the industry's position deteriorating in 1928, NSW Premier Bavin held a number of meetings to discuss the problems in the industry and to find possible solutions. The so-called "Bavin plan" emerged from these meetings, with the Premier proposing a cut in coal prices to restore the NSW industry's competitiveness. The employers, employees and the state Government would share the cost of the price reduction: the employers through the lower prices, employees through a cut in wage rates, and the Government through lower rail freight and port charges.[228] The Commonwealth Government also offered an export bounty as part of the plan.

In February 1929 the central council of the Miners' Federation rejected the plan, and instead called for the Commonwealth Government to appoint a Royal Commission to investigate the industry, with a key issue to include the profitability of the coal companies. The northern district employers gave 14 days' notice of dismissal to their employees. The employers were intent on only re-opening the mines if the mine workers were prepared to accept wage reductions. No agreement on wages was forthcoming from the miners, and on 2 March the northern mines were closed; the employees were locked out.

With the lockout now a reality, on 22 March J&A Brown's boss, John Brown, was singled out by the Commonwealth Government and prosecuted under the Industrial Peace Act of 1920 and the Industrial Conciliation and Arbitration Act for the lockout at his Pelaw Main and Richmond Main collieries. However, following a meeting in April with the NSW Premier Bavin and representatives of the miners and owners, Prime Minister Bruce announced that the charges against Brown would be withdrawn. That meeting had been convened to consider how to get the industry back to work, but Bruce said that he had told the meeting that it would not be possible to proceed with the meeting if the prosecution was still proceeding. Bruce claimed that no-one at the meeting disagreed.

The Opposition later in the year moved a censure motion in Federal Parliament slamming the Government: "That by its withdrawal of the lock-out prosecution against the wealthy colliery proprietor John Brown after its vigorous prosecutions of trade unionists, the Government has shown that in the administration of the law it unjustly discriminates between the rich and the poor..." Bavin justified his action saying that experience had shown that penalties did not prevent strikes or lock-outs, as had been evidenced by recent prosecutions of the Waterside Workers federation and the Timber Workers Union. Equally, he said, the threat to prosecute Brown had not brought about a resumption of work. John Brown was undeterred by the Government's actions and in May released a statement saying that he "most definitely" contradicted rumours that an early re-opening of his mines was likely. Brown stated "emphatically" that there was no truth in articles in some newspapers that he was considering re-opening his mines. He reiterated that there had to be sacrifices by all parties – "government, owners and men" – and that "his mines would not re-open unless wages and other charges were reduced and the industry was placed on a proper footing." [229]

The lockout dragged on, and in December 1929, one of the most violent demonstrations in Australian industrial history took place outside the Rothbury colliery near Cessnock. The colliery had been taken over by the NSW Government which was preparing to re-open the mine on 16 December using non-union labour. Several thousand miners marched on the colliery overnight, and a violent clash occurred between the demonstrating miners and the large contingent of police guarding the colliery. One miner, Norman Brown, was hit by a bullet and killed, and many other miners were injured, as were a number of police. The Rothbury Riot, as it has come to be known, was one of the most infamous incidents in Australian industrial history, with blame being placed on the NSW Government, the police and the coal producers on the one hand, and the miners and their leaders on the other. The coalfields in the north were never the same again.

The month of December also saw the Commonwealth Government attempt to broker a settlement of the lockout through a compulsory conference chaired by Justice George Beeby of the Arbitration Court. Beeby recommended a re-opening of the mines on the basis of wage rates existing before the lockout, and with the selling price of coal to be reduced by 40 cents per ton, with the State Government contributing 20 cents, the Federal Government 10 cents and the coal producers absorbing 10 cents. Beeby also proposed that the dispute would be referred to the Court for a full settlement. The producers rejected the proposals.

The situation in communities in the Newcastle region was dire. "Many people unable to afford rent or house repayments moved to illegal shanty villages, some went 'on the wallaby'; some set their houses alight to collect insurance; many moved to tent communities beside Lake Macquarie or Nelson Bay, where they could fish for their dinner. Those remaining virtually turned Kurri into a market garden. Locked out workers and their dependents survived on State and Miners' Federation relief payments that provided ($4.20) a week to a family of four at a time when this was half the minimum living wage."[230]

In March 1930, the NSW Government ceased paying unemployment benefits to miners in the northern coalfields. With miners and their families starving, the outlook was grim. But by the end of April the craft unions had decided to get their members to return to work, and on 3 May the central council of the Miners' Federation decided to recommend a return to work. Mass meetings of mine workers accepted the recommendation and the lockout ended in June 1930, 15 months after it began. The collieries re-opened, with contract wages reduced by 12.5%. The mine workers had gained nothing. Many of the mine workers were never re-employed, or would wait years to secure another job in coal mining. In 1940, there were still almost 4,000 men registered for employment in the NSW coal mining centres.

Davidson Royal Commission

With no end in sight to the lockout, in May 1929 the NSW and Commonwealth governments appointed a Royal Commission into the NSW coal industry, chaired by Justice Colin Davidson, a judge of the NSW Supreme Court. The terms of reference for the Commission were wide and included the production, carriage, export, distribution and sale of coal, and the factors which led to the current situation in the industry. They also included the distribution of powers between the Commonwealth and the States in relation to the industry. In June the acting NSW Premier also asked the Commission to urgently investigate the profitability of the major northern district coal companies for the year 1927. The Commission reported back to the Government in late September, and to the angst of the Miners' Federation, the Commission found that the average profit of these producers was the equivalent of around 22 cents per ton. The average selling price in 1927 for all NSW coal was around $1.76 per ton.[231]

The Commission said that the objective of the acting Premier's request in June was to establish the profitability of the mines so that an assessment could be made of what contribution the producers could make in terms of a reduction in the selling price. It found that this contribution was around 6 cents per ton. The Federation rejected the Commission's findings, claiming that the average profit was much higher than the Commission had estimated. It believed that the Commission's calculations did not properly account for some of the costs and therefore seriously understated the real profit position. Premier Bavin welcomed the Commission's report, stating that it supported what he had said to the NSW Parliament earlier in the year when he proposed his industry plan, that is, that the average profit was around 20 cents a ton. Bavin also consulted with the employers who were prepared to re-open the mines on the basis of his plan, but the Federation was not prepared to agree.

The Commission's main hearings got underway in June and Charles McDonald, the chairman of the Northern Coal Association, was one of the first witnesses to give evidence. McDonald referred to the major reductions in sales to interstate and overseas markets which had occurred over the preceding few years, with exports from Newcastle falling from over 1.1 million tons in 1922 to 360,000 tons in 1928, and the total output of the northern district dropping by 25% from 8 million tons in 1924 to 6 million tons in 1928.[232] McDonald attributed the decline in the industry to a range of factors, including competition from South Africa and Japan, the use of fuel oil in shipping, the development of the electricity generation in Victoria based on brown coal, and the ongoing strikes which had led to consumers turning elsewhere for their coal. McDonald said that he had been told by the Premiers of both Victoria and South Australia that they would continue to import coal from overseas.

The Miners' Federation northern district secretary, David Davies, suggested to the Commission that the majority of the northern coal companies were earning satisfactory profits, and that they could continue to pay award wages and conditions. The Federation believed that the producers had been able to hide their profits by excessive depreciation allowances, and through selling arrangements under the interlocking relationships between the coal producers and the shipping companies.[233] Jack Lang, Premier in 1925 to 1927 and from 1930 to 1932, had a similar view: "By 1927 the northern mines had reached their peak of prosperity. They had a flourishing export trade…Profits were concealed in many ways. There was an interlocking with some of the interstate shipping companies. …The big profits were being made through subsidiaries and distribution. They all went into the same pockets."[234]

In 1929 the links between the northern NSW coal producers and the major shipping companies were still strong and almost 70% of northern coal production was controlled by a small number of coal companies with ownership or other strong links to the shipping companies. As the earlier Campbell Royal Commission had noted, Caledonian Collieries was linked to Australian Steamship Company through Howard Smith; Abermain Seaham Collieries was linked to Adelaide Steamship Company; Hebburn Collieries was controlled by Huddart Parker; Hetton Bellbird Collieries was linked to McIlwraith McEacharn; and J&A Brown continued to own and operate its own shipping fleet.[235]

The Davidson Royal Commission reported in 1930, and but for a change of Government in NSW in 1932, the course of the industry in the decades ahead is likely to have been quite different. The Commission found that there were two main causes of the industry's problems – over-capacity and the hostility between the employers and workers, or in its own words "the spirit of hostility between capital and labour".[236] It also concluded that it was inevitable that at some stage a number of mines would have to close and that coal prices were too high. Its report compared pit-head prices in the USA, Britain and Australia and found that the cost of coal was "undoubtedly excessive at this time and the industry must submit to a process of readjustment to ensure that coal will be available to make the price conform with its real value."[237]

One of the features of the Commission's report was its quantification of the extent of the industry's excess production capacity. Assuming the existing coal prices continued, the Commission estimated that the excess capacity was around 48% in the northern coalfield, 25% in the south, and 50% in the west. It also quantified the excess number of workers in the industry as 40% state-wide, with 42% in the north, 41% in the south, and almost 26% in the west. The total number of workers considered to be in excess of the number required was around 6,000, or around two in every five in the industry. On the basis of a reduction in coal prices to levels sufficient to restore the demand for NSW coal to its 1927 level, the Commission estimated that the number of workers was still 15% above what was required.

On the question of competition, it concluded that there was no evidence that competition was restricted for the export market or for the NSW market. It did however conclude that, based on the evidence of major coal users in other states, the shipping companies were able to dominate the distribution of coal in those states. The Commission in fact was fairly relaxed about the role of the shipping companies, saying that apart from the actual carriage of coal in ships, the power of the shipping companies was not such as to constitute a monopoly. However the Commission acknowledged that these companies did have a detrimental effect on the coal industry, not so much because of how they might restrain competition, as in their control over freight rates, and in some cases the fact that they were responsible for introducing practices into the coal mining industry which undermined the control of management because of "shipping expediency".[238]

Following its report to the Premier in September 1928 on the profits of major northern producers, the Commission's final report presented data for the State and for each district for 1927 and 1928. The ratio of profit to capital employed for the State was 9.6% in 1927, falling to 6.7% in 1928. The district figures for the two years were: north 8.1% and 6.3%; south 14% and 9.1%; west 12.1% and 5.1%. Contrary to the claims of the Federation and politicians such as Jack Lang, the Commission found that "whatever may have been the position in earlier years" the returns in the majority of cases were not unduly high, either in terms of profit per ton, or in terms of the return on capital invested.[239]

One of the important pieces of work undertaken by the Commission was an analysis of wages in the industry which showed a wide disparity in the earnings across the industry. The Commission surveyed producers and obtained data covering about 85% of employees in NSW for the financial year 1927-28. Despite the fact that the average number of days worked per week was less than 3.5, the average wage for the coal industry (around $12 per week in the north and the west and around $9 per week in the south) was still above the average for all industries and well above the basic wage. However around one in five coal industry workers earned less than the basic wage, and around one in two less than the all industry average. There were also a number of coal industry workers who were doing much better than average, with one in twenty earning more than 70% above the all industry average, and a few earning at least double the all industry average.[240] The Commission's report also showed a breakdown of the wages for over 2200 workers employed on a contract basis in Greta seam mines for the 1927-28 year. The median wage for these workers was around $14 per week. Around one in twenty of the Greta miners however earned less than around $8.30 per week, and almost one in four earned less than around $10 per week. Almost one in five on the other hand earned more than around $19 per week.[241] This was a wage structure which reflected the way in which the industry had developed, including with the contract system, the many different arrangements between mines and the varying geological conditions across the industry. Proposals for across the board wage reductions which were being pushed by the producers and the NSW Government in 1928 and 1929 therefore would have reduced earnings for those at the top of the scale who may have been able to cope with lower wages (mainly contract miners). However cuts would also have hit for the many workers earning below the coal industry average, particularly the one in twenty earning less than the basic wage.

The Commission's solution to the industry's problems was a new organisation with a board of three members, those members not being from the coal industry. "Selection of its members from the coal industry where suspicion is so pronounced would be fatal to success. It would be equally inadvisable to make a choice from political motives or from amongst the numbers of persons who have inquired into the affairs of the industry or presided over its many tribunals. The endeavour should be to procure the services of a sound chairman of sound training and recognised tactfulness."[242] The new board would have wide-ranging powers and functions including fixing selling prices, compiling detailed statistics on the industry, fixing wages and conditions of employment, arbitration of   disputes, determination of which mines should be closed, control of the allocation of coal, issuing of mining licences, and assigning quotas for all districts and mines. Davidson would again come into prominence just after World War Two in 1946 when his report to the Federal Government on a detailed inquiry into the coal industry, would again recommend a new organisation to reform the industry, led by a board of independent directors without links to coal employers or unions.

The Lang Labor Government was returned to power at the October 1930 election, and by November the Government's plans for the industry were starting to take shape. In November Mines Minister Baddeley announced that he had been conferring with the industry and was considering putting to Cabinet for approval a plan which would involve establishing a new board or tribunal to close uneconomic mines, force the merger of coal companies, introduce more efficient methods of production into the industry, fix prices, and settle disputes.[243]

A Coal Bill was introduced into Parliament in August 1931, and was subject to extensive debate and amendments. The Bill gave the proposed Coal Industry Board comprehensive powers to control the coal industry in NSW, with the Board of three members (one representing coal mine owners, one representing employees, and the third appointed by the Government as chairman). The powers would be along the lines recommended by Davidson, and included setting maximum sales for each district, setting the production for each mine, fixing the profits of each mine, and controlling the allocation of coal for various uses, including the compulsory acquisition of coal. The Board was also to have power to set the conditions under which mines could operate, the power to operate any mine, the power to close mines and to "aid in the settling of industrial disputes or in preventing threatening, impending, or probable disputes."

By 13 May 1932, the Lower House had agreed to accept the final amendments to the Bill passed by the Upper House the day before.

However, that very day was the day that the NSW Governor, Sir Philip Game, sacked the Lang Labor Government. An election was then held in June and Labor lost power. Under the new Government the Coal Bill lapsed, and with it the plans for the coal board and the reform of the industry. Whether Baddeley's Coal Industry Board would have been successful in improving the performance of the industry can never be known, but in any case the 1930s would prove to be the lost years, characterised by years of massive job losses, unprofitable mines, intermittent work, and lack of reform of the industry.

Great Depression devastates the nation

The Great Depression hit countries all over the world and Australia was no exception. Our national unemployment rate was around 7% in 1926 and 1927, but rose to around 11% in 1928 and 1929, before surging to 19% in 1930, the first full year of the Depression years. Unemployment continued to rise to 29% in 1932, before easing to around 20% by 1934.[244] Wages fell sharply during the Depression throughout the country, and over the period from December 1930 to December 1933 the average wage was down by around 20%.

The Depression in Australia in a technical sense lasted until 1934, when GDP recovered to the 1929 level, but its effects dragged on throughout the 1930s. The coal industry was hit harder than most. In NSW coal production plummeted by around 45% from the peak in 1926 and 1927 to the low point in 1933. Employment in NSW coal mining, which had reached a peak in 1926 and 1927 at around 25,500, fell to around 21,500 in 1928. The Depression and the impact of the northern lockout saw numbers fall to around 14,500 in 1930, and then drop further to 13,300 in 1933. Employment numbers remained at the 1933 level for the next two years, before the start of a recovery in 1936. Queensland's coal industry also suffered severely. Production in Queensland peaked in 1929 at just under 1.4 million tons, and fell to around 840,000 tons by 1931 and 1932. The reduction of close to 40% in production however was not matched by such a severe fall in employment, with jobs in the Queensland industry dropping only around 12% from almost 2800 in

1929 and 1930 to just under 2500 by 1933. However these figures do not reflect any cut backs in the number of hours or days for which mineworkers were being paid. This intermittency plagued the industry in the 1930s in particular and we will look more closely at the problem later in this chapter.

The coal companies too were severely affected during the Depression and the years that followed. Many sold coal at prices which did not cover their costs. An analysis of the accounts of 9 public companies in 1950 which accounted for around one third of NSW coal production found that 8 of these paid dividends in 1928 averaging 8% of paid up capital; 7 of these companies paid out between 10 and 15%. But by 1930 the dividend rate had fallen to 1.5% on average, and by 1936 to 0.5%, with some companies paying no dividend at all. By 1939, while production by these companies had regained the levels of 1927, their return on capital was still only 3.5%.[245]

All the coal producing districts suffered during the Depression. In Newcastle, the combination of the lockout and the Depression was particularly severe. Just prior to the resumption of work in the collieries, the president of the local chamber of Commerce said that coming at the same time as the general depression, the trouble had been little short of disastrous for the city.[246] Since the mid-1920s, the NSW industry had seen a major reduction in its interstate and overseas markets. In 1924, around 5.4 million tons, or around 47% of NSW production had gone interstate or overseas, but by 1930 that figure had slumped to less than 2 million tons, or 27% of production. There were many factors behind the loss of these markets including the rise in coal prices, the increasing switch to oil by ships, development of coal mining in other states, and the impact of the Depression. Before the First World War, NSW coal had been significantly cheaper than coal in countries such as the USA and Britain. By the late 1920s, the situation had been reversed, and NSW coal had become uncompetitive.

When the Depression hit, the coal industry in Queensland was also struggling. Production from private mines was down in most districts. In his annual report in September 1929, the president of the Queensland

Colliery Proprietors' Council, J F Walker, said that the combination of the lower production and greater number of collieries in the West Moreton district had created an unparalleled position in the industry. In 1928 production in the West Moreton was lower than in 1920, and at least 10 new collieries were operating.[247]   Walker also referred to widespread losses being sustained by the non-government producers in the industry. However on a positive note, Walker also compared the industrial disputes record of Queensland coal mining with that in NSW, saying that no major strikes or stoppages had occurred in Queensland since the general stoppage in 1916. NSW on the other hand had suffered 4,500 strikes, involving a loss of 8 million working days between 1914 and 1928, he said, adding that his State could be proud of the service it had provided to Queensland coal consumers.

Queensland moves to regulate the industry

The 1920s had been a reasonable period for the Queensland coal industry, with many of the producers being members of the Proprietors' Council. While there were also many producers who were not members, the Council was a sufficiently strong organisation for it to play a major role in determining how the industry was run. The West Moreton producers actually entered into what Davidson described as a "trade agreement" under which they undertook to refrain from poaching each other's customers. The producers were happy to have such an arrangement so that there was a degree of stability in the industry, and an absence of cut throat competition.[248]

Davidson, in his 1946 Inquiry report, concluded that when the trade agreement was operating, it largely achieved its aims in a period when the market for coal was either relatively steady or rising somewhat. Producers who were parties to the agreement saw it as a "stepping stone to something more permanent and whose provisions could be enforced - one of the obvious weaknesses of any voluntary agreement."[249] Those producers, according to Davidson, who had been in business for many years, and who had experienced the ups and downs of the industry, were working towards a goal, "nebulous and undefined at that time – whose attainment would enable them to produce and sell all their coal in an orderly and sensible way." He said that "the most important result of this striving for something better was the gradual growth of the conviction that a great deal could be done if the industry were organised" and that individualism was an outdated concept in a world which would only listen to groups which were organised. But then came the Great Depression and, in the years following 1929 (which was a better year for Queensland producers who benefited from the northern lockout in NSW) production fell across the State. In the West Moreton, production dropped by around one third between 1929 and 1933. With the downturn in the industry a number of producers openly rejected the agreement and resigned from the Council. "Those who remained in the Council were faced with the gloomy prospect of watching the industry slowly bleeding to death."[250]

By 1933 the state of the industry in Queensland was still grim, with price competition rife and many small mines having opened to mine coal from near the surface. These small mines did not have the intention or the finance to develop deeper operations and so be sustainable longer term, but their costs were low, and the larger collieries were suffering. The Mines Department told Mines Minister Stopford that there was little or no development underway at the established mines, and the producers said that they were unable to justify spending scarce funds on mine development or on exploration. The mine workers were afraid for their future with the uncontrolled loss of customers by mines to competitors and the absence of development work.[251]

The conditions were now ideal for the industry to get together to consider a solution to the industry's woes. The Minister convened a conference of industry leaders in June 1933 which agreed to ask the Government to legislate to establish a central coal board and district boards which could control the industry. A committee was then appointed to draft detailed proposals for consideration. The producers played the leading role in the committee, with the union, the Queensland Colliery Employees' Association, also involved. The committee of course was able to draw on experiences in other countries, including Britain, and on Davidson's Royal Commission into the NSW coal industry in 1929-

30 and the NSW Coal Bill which almost became law in 1932. A second conference in July endorsed the committee recommendations and by October 1933 the Coal Production and Regulation Bill was tabled in Parliament. The Bill, modelled largely on British legislation, was passed in December and by February 1934 the new Act was in force.

In tabling the Bill, Stopford stated that "a large number of collieries, working without capital, had taken out the coal from near the surface, and when they reached the stage at which development work was necessary they ceased operations. These small collieries were worked by parties of miners; they had no overhead expenses, and were able to avoid the expenses incurred by the large established collieries in complying with the coal mining regulations. As a result these small collieries could effectively enter into cut-throat competition with the established collieries." Stopford contended that this Bill would end that unfair competition.[252] The Act provided for a Central Coal Board and district Coal Boards to be established. Each district board was responsible for developing a plan for that district, with each plan needing the approval of at least 75% of the industry, with voting heavily weighted in favour of the larger collieries. These boards had the power to issue new licences for new mines to open, to allocate production quotas for each mine, and to set the prices at which coal could be sold.

The Leader of the Opposition went on record as saying that all the big collieries would give their unqualified support to the proposal. "It was practically a form of subsidy to the coal industry" he said, "which was given power to fix a minimum price for its product. This was a new form of price-fixing. Usually a maximum price was fixed to protect the interests of the consumers. Existing large collieries could prevent the establishment of new collieries able to produce coal more economically than they could produce it. The minimum price also removed the incentive to efficiency in existing mines. He could see only two results from the operation of the Bill - the squeezing out of many of the smaller collieries and the raising of the of the raising of the price of coal. Queensland could not produce coal at a price that would allow it to compete successfully in overseas markets."[253] While the Opposition

Leader was right about Queensland being unable to compete to secure overseas markets, his prediction about prices was not borne out by the events of the next few years. The average value per ton of coal in 1934 and 1935 was only marginally up on the 1933 level, and only rose slowly over the rest of the decade. The average value in 1940 was about 12% up on the 1933 level.[254]

However, while the new system may have stabilised the industry in Queensland in the post-Depression years, there were some adverse reactions from coal consumers. The Courier Mail ran a series of articles in 1936 which were critical of various aspects of the Act and the ways in which it was being administered by the Central Coal Board and the district boards.[255] It also reported on criticism from coal users on the prices they were forced to pay and the sources of supply which the Board directed them to use. The newspaper acknowledged that the operation of the Act was a great benefit to what had been a dying industry, and had "brought new hope to coal producers in the West Moreton district." It also said that 75% of the mines in the West Moreton had been unprofitable for the previous 8 or 9 years, and that some had paid small dividends of 1 or 2% after making large capital outlays.

The Brisbane Gas Company came out publicly to attack the Central Coal Board, its chairman, Mr T Cowlishaw, saying that under the legislation the coal-using public "must really come last and be wide open to exploitation."[256] Cowlishaw went on to say that the operation of the various boards had resulted in increased prices of coal, and had given colliery owners a monopoly in terms of the supply of coal to individual customers allotted to them by the boards, irrespective of whether their coal was the most suitable for that user. "Purchasers of coal are denied the right of selection, and have no direct representation either on the district coal board or on the Central Coal Board, to which latter board appeals are made…(The) Act is utterly one-sided in construction, giving no thought to the coal consumer, and ignoring his right to have the kind or quality of coal best suited to his particular purposes. In effect, the selection of the coal which purchasers are to use is transferred to the coal seller, who has no knowledge of the purchaser's requirements. This state of affairs must ultimately result in detriment to the coal industry. Public utility undertakings are obliged by legislation to provide a good service at a reasonable cost to the community, and nothing should intervene to lessen their capacity to do that effectively and at minimum costs. They should therefore be free to have the quality of coal they desire. They should be the only arbiter of the quality best suited for economical gas production."

The Courier Mail accused the Central Coal Board of acting like a dictator, with a major story headlined "Economic Facism in Coal Industry." In response to price increases approved by the West Moreton Board, the newspaper said that when the legislation establishing the Boards was being debated in Parliament in 1933, the Minister "was at pains to allay any fears members might have (had)", with the Minister saying that the real object of the legislation was "not to Increase the price of coal to the consumer at all, but to secure the proper working of a national asset. I have watched the Interests of the consumer. I have prevented any power in this Bill which might be used detrimentally to any section of the community."[257] The newspaper also criticised the make-up of the board responsible for the major coalfield, the West Moreton, which was dominated by colliery representatives. Under the Act, the district boards had a board of directors of up to seven, one of whom was an employee representative, with the Government also able to nominate one director and the colliery proprietors taking the other positions. The West Moreton Coal Board hit back at the Courier Mail articles with a lengthy statement which was read out at a public meeting in Ipswich by its chairman and secretary. "Shareholders in gas and electricity undertakings are guaranteed a return upon their investment" it said, adding that "depositors in the Commonwealth Savings Bank are guaranteed a return of 2 per cent and surely no one will contend that the coal industry — the life-blood of all industries— is not entitled to sell its product at a price which will return a profit commensurate with the risks associated with all colliery investments." [258] The statement went on to say that a number of coal consumers including the gas industry appeared "to have lost sight of the fact that the selling price of coal must return a profit to the coal owners if the industry is to maintain a regular supply of coal from properly developed collieries." It also went into details, pointing out that the 6 collieries supplying coal to the Brisbane City Council and the Brisbane Gas Company lost money on every ton of coal produced in the twelve months to June 1934, a fact which it said would be "staggering to those who are under the impression that enormous profits are being made by coal owners..." The data comparing production costs with selling prices had been prepared by a firm of accountants, with the Board saying that the losses did not include allowances for depreciation on railway sidings or expenditure on shafts and tunnels, or uninsurable risks such as floods, fires or other contingencies peculiar to coal mining. With wage increases since the data was complied, the Board also explained that "practically every purchaser of slack coal, including the Brisbane City Council, the City Electric Light Company, and our secondary Industries, are still buying coal below the cost of production."

In another major article attacking the system, the Courier Mail called for change, pointing out that coal prices had been increased significantly to the Brisbane City Council and other major users. The Council it said was using around 60,000 tons per year, with the Brisbane Gas company around 40,000 tons, but in common with manufacturing industries, they had been denied obtaining the coal most suitable for their requirements.[259]

There was clearly bad blood between the Boards and the major coal consumers which saw the industry as overcharging for its product and denying them the appropriate qualities of coal they required. While the establishment and operation of the Central and District Boards may have assisted to stabilise the industry, the new system also placed the industry in a cocoon, with producers effectively regulating themselves. The boards continued to operate until 1948, when they were replaced by the Queensland Coal Board, the Queensland Government's answer to the Joint Coal Board set up in NSW in 1947. The era of the coal boards in Queensland would come under intense scrutiny in 1948 when the Queensland Government commissioned expert British consultants to examine the industry and recommend the way forward. The experts' report would prove to be an indictment of the industry which had failed over many years to mechanise and prepare for the growth in demand for coal which was to occur in Queensland after the end of World War Two. Davidson, in his 1946 report, also would have some harsh things to say about the system which controlled the industry from 1934 and through the war years.

The union war to oppose mechanisation fails

BHP's Newcastle steelworks began operating in 1915 and for the early years of its life the company bought coal supplies from collieries owned by other companies. But by 1923, after complaining for years about the unreliable supply and the high cost of coal, BHP decided to take things into its own hands. The company's chief, Essington Lewis, then with the modest title of chief general manager, announced in 1923: "The experience of our officers, who have visited steel plants in other parts of the world has been that wherever possible steelworks of any consequence have acquired their own coalfields… In the light of this experience it was decided during the early part of the year that the company should investigate the possibilities of procuring its own collieries…Options have been secured over large tracts of land, which are now under investigation.'[260]

BHP bought almost 4000 acres of land in the Maitland district and, with Hebburn Collieries as the only other shareholder, formed a new subsidiary called BHP Collieries Pty Ltd. Its next major steps were the development of a new colliery called Elrington near Cessnock, and the purchase and development of an existing colliery south of Newcastle near Belmont, which it renamed John Darling. Development of Elrington commenced in 1924 and the first trainload of coal was despatched in 1928. Development of John Darling began in 1925, with the first trainload of coal despatched later that year, but because of industrial disputes, full production did not commence until 1928. These collieries fed the company's steelworks in Newcastle.

In 1932 the company moved further into coal mining with the purchase of the Burwood and Lambton collieries near Newcastle from the Scottish Australian Mining Company. This was followed in 1935 by BHP's purchase of Australian Iron and Steel (AI&S). Gaining control of AI&S not only brought the Port Kembla steelworks into the BHP camp, but also its substantial colliery assets in the western and south coast districts – Lithgow Colliery in the west and the Bulli, Wongawilli and Osborne Wallsend collieries in the south. BHP had now become a major force in the industry, but it was not only its size that was to have a major impact on the industry, but also its plans for mechanisation. The Joint Coal Board would become a driving force for the mechanisation of the industry in NSW after World War Two, but BHP was the driving force in the 1930s. And BHP would pay a high price for taking the leadership role in terms of industrial disputes for many years.

The start of mechanisation, at least as far as BHP was concerned, was 1934, with the introduction of Jeffrey coal cutting machines from the USA into the Burwood colliery. Next came Lambton colliery which the company closed in 1935 to install coal cutters, loaders and electric battery locomotives. Lambton was the first coal mine in Australia to mechanise the loading process. Prior to the installation of Jeffrey loaders, loading at Lambton had all been by hand, with workers using forks or shovels to fill skips which were then taken to assembly points in the mine, before the coal was sent to the surface.[261] BHP claims that towards the end of 1935 Lambton was the first colliery to produce coal which was 100% mechanically mined and loaded.[262] When BHP re-opened Lambton, all the coal mined at the coal faces was cut by electric coal cutting machines, three mechanical loaders were in use in the mine, and electric locomotives hauled the coal from the face to the shaft from where it was sent to the surface.[263]

In 1936, the Wallarah Coal Company announced that it was also planning to completely mechanise its Wallarah Colliery. That year it ordered six electrically powered Joy mechanical loaders and by the end of the year it had received delivery of two of the loaders.[264] BHP's John Darling colliery had a coal loader installed in one area of the mine in 1937, and the AI&S south coast mines then followed, with coal cutters in Wongawilli in 1938 and complete mechanisation of sections of Bulli and Mt Keira in 1940.[265]

Following the re-opening of Lambton in 1935, the leaders of the Miners' Federation in the northern district tried to block the mine from continuing to operate. Lambton had started to produce coal using colliery officials (non-production employees who were not Federation members) operating a new mechanical loader. The company had not sought the agreement of the Federation or other mining unions, and the Federation's northern board of management issued an edict to its members that they should withdraw from the mine. The miners initially complied, but quickly returned to work and refused to resume the strike. "The general campaign against mechanisation fizzled out, although there was continued resistance, here and there, on particular aspects, such as safety and health."[266]

BHP's determination to mechanise had meant an early confrontation with the new guard at the Federation, and a win for the company. The company's determination to begin to mechanise its mines may also have been a factor in other companies making some tentative steps to also mechanise. For example, by 1938 mechanical loaders were in use not only in BHP's Lambton, Burwood and John Darling collieries, and in Wallarah, but also in JABAS' Abermain No.2 and Richmond Main, and in Bloomfield Main and Dudley collieries. Abermain No.2 colliery had been closed at short notice in mid-1936 throwing over 400 employees out of work; after lengthy negotiations with the unions, it re-opened late in 1938 substantially mechanised. By 1939, mechanical loaders were also in use in the NSW southern district at Osborne-Wallsend, Coalcliff and Nattai Bulli collieries, and other collieries in the northern district had also begun to use them, together with the Katoomba and Invincible collieries in the western district (Katoomba had turned to mechanical loaders in 1937).

The progress towards mechanisation in NSW in the 1930s is evident from the proportion of coal produced by mechanical coal cutters and mechanical loaders. In 1931, in the depths of the Depression, just over 21% of coal was produced from mechanical coal cutters (both electrically driven and compressed air driven). By 1939 the proportion had moved up to just over 32%, and by 1939 almost 10% of coal was loaded underground by mechanical loaders. The trend to mechanise this key part of the coal production process was certainly well underway by the time of the outbreak of the Second World War, but there was a long way to go before the industry could be described as technically up-to-date or largely mechanised.

The new guard at the Miners' Federation

The leadership of the Miners' Federation changed in 1934 with the election of William Orr as general secretary and Charles Nelson as president; Orr was regarded as the dominant one of the duo.[267] The son of a mining engineer, Orr had served in World War One before migrating from Scotland to Australia. Both Orr and Nelson were communists and criticised their predecessors for being too passive and for accepting the conditions in the industry. On taking office Orr and Nelson began a campaign to oppose mechanisation of the industry. They saw mechanisation as a threat to thousands of jobs and developed a strategy to promote their opposition to mechanisation and published pamphlets as a key part of that strategy, including *Mechanisation: Threatened Catastrophe for Coalfields* and *Coal: The Struggle of the Mineworkers*. Nelson continued as president until December 1941 when he was defeated for the position by Harold Wells, another communist. Orr served as general secretary until ill health forced him to retire from the position in 1940. The Federation's magazine for its members, *Common Cause*, was also revived in 1935, with Edgar Ross as its editor. Ross would continue as editor for 31 years, and the journal would become a major channel for the Federation and its propaganda. The Orr/Nelson-led Miners' Federation would see huge gains for its membership in the following few years, gains arguably unequalled in the union's history to that time.

Communist class warfare rhetoric was rife throughout the Orr and Nelson publications, for example with the Foreword to the first showing their desire to overthrow the prevailing system: 'Between these two classes (working people and the employing class) the struggle must continue until capitalism is abolished.'[268] And the Federation claimed that "Despite all the 'ballyhoo' about returning to prosperity" the conditions of the coalminers had not improved, but in fact had become worse.[269] They also pointed to what they saw as "anarchic, price-war competition" amongst the mine owners and exploitation of the mineworkers.

Following the drastic job losses during the Depression, the Federation now saw the decimation of its membership as a result of mechanisation: "The offensive against the mineworkers continues, and new moves are now being made by the coalowners in the direction of complete mechanisation; electric coal-cutters, loaders and locomotives threaten the scrapping of the major portion of coalminers in the next few years."[270] Orr and Nelson also referred to the impact of mechanisation on activities on the surface of mines: "Cases already exist where surface operations are almost completely mechanised to the point of dispensing with all but a few employees, and if the coalowners get their way the same development will be applied underground in a comparatively short time."[271]

In 1937 the Miners' Federation in NSW laid out a program of change it expected to see in relation to working conditions, wages, safety, pensions and legislation. The key demands in this program, which formed the basis for its campaign over the coming years, included: working hours to be reduced to 30 hours per week per week underground and 35 on the surface; a minimum wage of 25 shillings ($2.50) per day for underground workers and 21 shillings ($2.10) for surface workers, with restoration of the 12.5% cut for all contract workers which had been made in 1930; creation of a permanent Safety in Mines Bureau to enforce new and safer methods of working; amendment of regulations and special rules to ensure a maximum of safety and health protection; establishment of a pensions fund for retired workers; and a new coal Act to control the production, prices and marketing of coal.[272]

This ambitious program and the Federation's 1938 log of claims laid the foundation for the gains which the Federation and the mine workers were to see over the coming years. Coming as they did after such a traumatic and turbulent period, the Federation's demands were remarkably successful, but unfortunately did not also lead to industrial peace or a more efficient industry in the following decade. Orr and Nelson

also signalled a challenge to other coal mining unions under the guise
of "trade union unity", claiming that this unity had been a consistent
policy of the Federation, and that amongst the smaller unions there was
a "growing urge" for the Federation to take over all the members of
these unions.[273]

The curse of intermittency

The Australian coal industry in the years before World War Two saw long
periods when mines would operate on an intermittent basis, sometimes
producing on only two or three days a week, and sometimes on only
three or four days. The periods when most mines were humming along
without interruption for five or more days a week were the exception.
Until 1939, the maximum number of days per year that a NSW colliery
could work was about 274, or about 5.5 per week, assuming that mines
closed down for a week or two over Christmas and New Year. In 1939,
changes by the Arbitration Court reduced the maximum to 266 days (the
changes applying for only part of the year). In 1940, when these changes
were in effect for the full year, the maximum was down to 244 days, or
just under 5 days a week. For the rest of the war years, the maximum
average around 250 days, or 5 per week.[274]

In 1923, one of the better years for the industry, NSW collieries
worked an average of 223 days, or around 81% of the maximum possible,
equivalent to an average of around 4.5 days per week. Later in the decade,
as the industry fell into a slump, but before the northern lockout and the
start of the Depression, the average days worked dropped to 187 in 1927
and to 168 in 1928. The numbers for 1929 and 1930 of course were
impacted dramatically by the northern lockout, but 1931 saw the real
impact of the Depression and the serious drop in the demand for coal.
In 1931 the average days fell to 141, or less than 3 days per week. The
next few years saw a slow improvement, and by 1935 the average for the
year had climbed back to just over 200, or around 4 days a week. The
late 1930s, except for 1938 (which featured prolonged strike action) also
averaged around 4 days a week or a little higher. The years during the war,
except 1940 (again a year seriously affected by disputes) would see the

average week at between 4 and 4.5 days. 1942 was the best year, with 231 days worked on average, or a little over 4.5 per week.

Comparable data for Queensland collieries was not published for the pre-war years, however it is possible to calculate the average number of shifts per underground worker from Queensland Mines Department annual reports. The calculations give us a similar picture to the situation in NSW. In the depths of the Depression for example in 1933, the Queensland coal industry's average number of shifts per underground worker for that year was 158, or around 3 shifts per week, with the Ipswich district averaging around 3.6 shifts per worker per week, and the other coal districts only 2.6 shifts per week. These calculations are based on data for the total number of shifts worked underground for the year, and number of men employed underground.[275] By 1935, similar calculations show that the situation in Queensland was improving, with an average of around 3.6 shifts per week underground. By 1939 the average had crept up to around 3.7 per week.

The Queensland Government Mining Journal in July 1933 reported no real improvement in conditions in the industry compared with 1932, with "unemployment and intermittent work very much in evidence."[276] The Blair Athol mines had only worked an average of one day a week in the first half of the year. The Bowen mines, including the Bowen State mine, had been working an average of less than half time, although there had been an improvement in the previous two months. The Journal in July 1935 reported a slight overall improvement in conditions in the industry, with the Ipswich and Rosewood districts seeing increased production and more regular work. However on the Darling Downs, two or three days work per week was the norm at almost all of the mines except the Federal Colliery. The Blair Athol mines on the other hand were experiencing better times by 1935, working on an almost full time basis.[277]

In an industrial case before the Southern Local Coal Board in February 1931, the Queensland Colliery Employees' Union (QCEU) representative said that the average number of days per week being worked in the industry was 3.[278] An even worse picture was given in

September 1935 at a hearing before the State's Industrial Court for a new coal award when the QCEU tabled a summary of wages and conditions which showed that Queensland miners had averaged only 2.5 days per week from 1932 to 1934. [279]

For the Queensland industry, the situation for coal miners and their communities in the early 1930s was grim. Employment numbers had dropped from the peak in 1929 and those who were able to hold on to their jobs were working only part-time. Some workers were lucky to average two days a week depending on where their colliery was located and how hard hit its customers had been by the Depression.

National strike of 1938

In August 1938 the Federation served a log of claims on the coal producers and on the NSW and Commonwealth Governments. The log was based on the program laid out in 1937, but was more specific on some demands. The key elements of the log were: a 5 day week of 6 hours per day, with no reduction in pay; all workers on piece work or contract work to be guaranteed a minimum wage; the introduction of a special Compensation Act for the coal mining industry; all mine workers on reaching 60 to be pensioned off with a payment of £2 ($4) per week; all employees to be paid for 14 days holiday a year; and all wages to be paid weekly. Similar claims were made by the other unions -- the FEDFA, the AEU, the MMA and Blacksmiths' Society which covered the engine drivers.[280] The log was rejected by the Arbitration Court and a general strike followed. The NSW Government then moved to allay the concerns of the miners by appointing two Royal Commissions – one into mine safety and a second to investigate the case for a pension scheme for retired miners.

Judge Drake- Brockman of the Commonwealth Arbitration Court was appointed to review wages and conditions in the industry, and as part of the case he conducted a detailed review of the industry which included a survey of the producers. He slammed the producers for the way in which they ran their businesses and concluded that there had been over-expansion in production, which together with cut-throat competition, had made the industry unprofitable. Drake- Brockman brought down an initial decision on the coal award in June 1939, with the full bench of the Court delivering a final award in October 1940. While the process from the initial wish list of the Federation in 1937, to the interim award in 1939 and the final award in 1940 was a drawn out one, the result was a win for the union. The underground workers had been granted a 40 hour week (over 5 days of 8 hours, bank to bank), and paid annual leave of 10 days for workers on the 40 hour week and 11 days for those working over 40 hours. Surface workers were unsuccessful in getting a shorter working week, and had to accept a continuation of the existing 86 hours per fortnight arrangement. Wages overall were not increased, although increases were granted in some cases.

The Miners' Federation and the other unions had had a major win with the Drake- Brockman decision, but they wanted more, including a new industrial tribunal for the coal industry. The previous Coal Tribunal had become effectively defunct on Hibble's resignation in 1930, and following the final decision by the Arbitration Court on the award, the combined mining unions asked the Federal Government for the special coal tribunal. Prime Minister Menzies said that Cabinet had rejected the request: "'The real dispute presented to Cabinet has been whether the jurisdiction of the Arbitration Court should, in effect, be set aside and some form of special tribunal set up for the coalmining industry. Cabinet is of the opinion that, under all circumstances, it is clear that the objection to the Arbitration Court is based upon dissatisfaction with its findings, and that the combined unions are seeking another tribunal in order that they may obtain from it some concession refused the Arbitration Court."[281] The unions were disappointed with the Cabinet decision and now looked to direct action to campaign for a 40 hour week for all mine workers and higher wages. The employers did not want a return to a special coal tribunal and were happy to remain under the jurisdiction of the Arbitration Court.

The unions now had to live with the Arbitration Court, but it would not be for long. The war years would see a completely new regulatory

structure for the coal industry, one of the key elements of which would
be a new coal tribunal, the Central Reference Board which was established
in February 1941.

Royal Commission on Safety and Health

The Royal Commission appointed in 1938 to investigate the safety and
health of coal mine workers and the need for legislative change was
chaired by Colin Davidson, the judge who had also chaired the Royal
Commission in 1929-30 into the NSW industry during the infamous
northern lockout. In the 1930s, safety in the coal industry had continued
to be a major issue, with fatalities a regular occurrence, and serious
injuries also far too prevalent. The Commission's report presented data
on deaths in NSW coal mines, and in coal mines in Britain and the USA.
In the 10 years from 1928 to 1937 there had been 137 deaths in NSW,
over 9000 in Britain and over 12000 in the USA. As a ratio of deaths
to numbers employed, the average for that decade in NSW was 0.91 per
1000; for Britain the ratio was slightly higher at 1.08, and for the USA
2.35. The Commission found that 65% of deaths in NSW had been due
to falls of rock or coal from the roof or sides of underground workings,
a much higher percentage than in Britain (50%) and the USA (53%).

Dust levels in mines were another priority area for the Commission.
The Miners' Federation pushed the Commission for wide–ranging
improvements in safety, and was particularly strong on the issues of
dust and ventilation. Charles Nelson, the general president, pointed to
the number of miners with respiratory and lung diseases, and said that
he believed that more miners had died from breathing harmful material
than had been killed by explosions or fires. Nelson also referred to the
practice of stone dusting, which he said was supposed to not be harmful,
but some producers were not using limestone but cheaper alternatives
which were highly dangerous. He also referred to modern coal cutting
machinery creating excessive dust which was harmful when breathed in
by miners, and claimed that the producers had made no attempts to install
water sprays on machinery, with the only watering being "spasmodic
splashing" of water on the underground roads.

The Royal Commission reported in July 1939, its report coming only one month after the historic Drake-Brockman coal industry award decision. It recommended a wide range of changes and improvements, including a requirement for companies to supply their workers with protective equipment, stricter rules for the use of explosives, improvements in air circulation and a prohibition on the use of naked lights, with the phasing out of naked lights within one year. The Commission also made a range of recommendations on stone dusting which it concluded was necessary to safeguard against explosions caused by coal dust. The key recommendations were incorporated into legislation in 1941with changes to the NSW Coal Mines Regulation Act.

Queensland moved ahead with some modest changes to its mining safety legislation while the Royal Commission in NSW was still hearing evidence, with changes to the Coal Mining Act passed in September 1938. One of the most important was the requirement for mines employing more than 20 men to have a mine manager who held a first class mine manger's certificate; in mines with 8 to 20 men employed the manager needed a second class certificate; in mines with up to 8 men a deputy's certificate would be required by the manager. New departmental inspectors now also had to hold a first class mine manager's certificate. Other changes included a requirement for new check inspectors appointed by mine workers to hold a deputy's certificate or higher qualification (previously check inspectors needed no formal qualifications).

The Royal Commission led to a number of much needed and important changes to coal mining regulation, but it would be the mechanisation of mines after World War Two which would see a significant reduction in fatalities, along with major innovations such as roof bolting. Often neglected however until the 1980s and 1990s were the cultural factors in the industry which were also important causes of deaths and accidents. A few examples here may help to illustrate the some of the challenges which the industry faced if it was to see the dramatic improvements in safety that were needed.

In 1928 the managing director of the Redbank colliery in Ipswich, and two men who were part owners, entered the mine on a Monday morning carrying naked lights and were killed by a gas explosion. The mine's ventilation system had been turned off that weekend to save energy, and had only just been turned back on that morning. The mine was known to be gaseous and had seen explosions in the past. One witness at the subsequent inquiry also stated that an open box of matches had been found near one of the bodies. Had that day not been a holiday and had the colliery been operating with the normal contingent of workers, Redbank may have been another of the industry's major disasters, with many more than 3 lives lost.[282] The practice of shutting down the ventilation system of mines had been common in the years before the World War One. For example, the Queensland Mines Department's annual report for 1911 for example referred to this practice, saying that it was common in the Ipswich district for the ventilating furnaces and fans to be shut down immediately after the end of the day shift, and not to be re-started until the deputy arrived the following morning. The 1911 report also noted that this was a 'bad practice' and against the spirit of the Rules of the Act which specified that the supply of air was to be constantly generated, and that it rendered the colliery manager liable to prosecution.[283] The Department's 1912 report indicated that the new Mines Regulation Act which took effect in 1911 had given the Department's inspectors the necessary powers to direct collieries on ventilation, and so the bad practices noted in the 1911 report seem to have gone. The Redbank colliery in 1928 seemed to have relapsed back to the bad old days.

A common practice in the industry, still prevalent well after World War Two, was for workers to ride on empty skips as they were hauled back to the coalface. It certainly meant a faster trip back to the coal face, but it was extremely dangerous and led to many deaths and serious injuries from workers being crushed between the skip and mine walls or the coal face. The NSW Mines Department stated that there were three fatal accidents in 1937 related to underground haulage, one of which involved a shunter who was crushed between the front of an empty skip he was riding and the coal face. The chief inspector stated that "It is a dangerous practice for locomotive shunters to ride on the front of a skip or front end of a set of skips."[284]

The manager of the North Wallarah colliery near Newcastle, Richard Marks, was struck in a haulage tunnel in 1937 by a set of empty skips and died in hospital the same night. The Mines Department's inspector who investigated the accident said that it appeared that the manager's practice was to advise the haulage engine driver when he was entering the mine, but that on this occasion he failed to do so. The engine driver sent a batch of empty skips down the tunnel in which the manager was walking; they apparently were derailed and struck Mr Marks. The inspector who was assigned to that colliery, but was unavailable that day, noted that the accident may not have occurred had "Mr Marks not been afflicted by deafness."[285] So here we have a deaf mine manager who went underground on his own and without notifying the man controlling the haulage engine, a tragic combination of factors which should never have been allowed to occur.

Accidents and deaths from the incorrect use of explosives were also common throughout the industry. The NSW Department's annual report for 1938 stated that 9 men had been injured that year in explosive accidents. Regulations required the miner or shot firer firing the shot to ensure that he and everyone in the vicinity had taken proper shelter. In several cases that year, workers were hit by flying material around 40 to 60 metres from where the shot was fired, relying on the distance alone for their safety. The chief inspector noted that if the regulations had been followed correctly, all the accidents would have been avoided.[286] The NSW Department's annual reports for this period also listed the prosecutions of mine workers and managers by the Department or prosecutions of workers by management each year. Far too common were the cases where workers had been prosecuted for taking matches and cigarettes underground in contravention of the regulations. Other cases which raise questions of lack of common sense, laziness or simply disregard for safety include carrying explosives underground in containers which were not secure, with one case in 1937 involving a miner who was fined for carrying explosives into the mine in a paper bag.[287]

These examples are just a fraction of the cases that can be found in the departmental and other records of the industry. Regulations can go so far, but good management, common sense and modern technology would prove to be equally important in the future in bringing the industry's safety performance into line with best practice.

Companies go backwards

By 1938, and with the notable exception of BHP and some other companies, the industry had progressed little since the disasters of the Depression and the 1929 northern lockout. In fact, on the basis of the value of plant, equipment and machinery used underground in coal mines, on the surface, and for transport to wharves or rail stations, the NSW industry had gone backwards. NSW Mines Department data for NSW show that the total value of plant, machinery and equipment fell between 1928 and 1938 by over 7%, and this was despite the major investments that BHP/AI&S and other companies had made in upgrading their mines with mechanical loaders and other equipment. Clearly many other producers had done little or nothing to invest.[288]

Many producers were making losses in the 1930s and had little incentive to invest in new equipment or otherwise upgrade their production capacity. Drake- Brockman's examination of the accounts of 30 of the main collieries in NSW found that the producers had frequently been selling coal at below the cost of production, and that from 1931 to 1937, these 30 collieries in aggregate had seen a reduction in the value of their capital of over $1.6 million, a substantial amount for those times.[289] Drake- Brockman was scathing about the state of the industry in his judgment in 1939, and in one of the most famous statements ever made about the industry he said: "The history of the coal mining industry in Australia from its very inception may be described as an unbridled and unregulated contest between employers and employees without restraint and actuated only by the rules of the jungle."[290] Drake- Brockman also said that "employees had usurped the functions of management" and that managers and superintendents seemed to lack the courage to impose discipline. The "last on first off" rule had been established, meaning that no matter what a reasonably long-employed mine worker did, or how lazy or insubordinate or mischievous he might be, he was firmly entrenched in his job. Drake Brockman added that "Nothing short of the closing down of the mine, death or compensable illness could prise him out of it."[291]

The state of the industry and the problems in achieving change were evident from the proceedings of the Drake- Brockman case. In a telling session in January 1939, the superintendent of Caledonian Collieries, Nathaniel Clark, was questioned by the legal counsel for the proprietors on whether production could be maintained if a six hour day was granted to miners. Clark said it could only be done by an increased number of shifts or by employing more workers, and that in any case, the miners would not work two shifts. Drake- Brockman fired back, saying that "They do it at Wonthaggi" (the coal mine in Victoria). Clark responded that he was certain the miners would not do it on the Maitland coalfield, and that the feeling against working the afternoon shift was so bitter that the miners would not be telling the truth if they said they would work a second shift. Drake-Brockman said he thought the thousands of men who had been retrenched would welcome the right to work a second shift, but Clark countered that they would not be allowed to do so by the strength of the Miners' Federation.[292] Drake- Brockman said he could not understand the defeatist attitude of the employers. So long as they took it lying down, he said, the coal industry would continue to be in a mess, adding that the owners complained about a lot of things, but in his opinion they did nothing to correct them. "You take no steps to have the industry put on a proper basis. Why on earth you don't do something about it is beyond my comprehension" he said.

Drake- Brockman's frustration was understandable, but attitudes in the industry on the part of both employers and employees were so entrenched that change would not come readily. But the war was about to start and Australia would be heavily involved. Change would come to revolutionalise the industry, but would have to wait until the late 1940s, when, after the defeat of Nazi Germany and Japan, a new era for the coal industry in both NSW and Queensland would begin. When the Second World War broke out in 1939, the Australian coal industry was in general in a poor state to face the challenges that the war would throw up for one of our key industries supplying essential coal to steel mills, gas making plants, railways, power stations and a wide range of manufacturing and other industries. Apart from the BHP/AI&S group and a few other producers, the industry was dominated by producers who were only partially mechanised or with little or no mechanisation at all. Industrial relations were caught in the vicious contest that Drake-Brockman had described. Safety standards were still poor, and deaths frequent.

4

World War Two: the industry fails to meet the challenge

In the late 1920s the Australian coal industry employed over 30,000 people, but following the Depression massive cutbacks saw the numbers down to less than 19,000 by 1935. By 1938 the industry had recovered somewhat, with employment up to 21,000. Around three quarters of the workers were in the NSW coal fields. The industry in Queensland employed around 2500, with most operations located in the Ipswich area west of Brisbane, in Central Queensland and the Darling Downs. Victoria had the Wonthaggi colliery in the south Gippsland region employing around 1300. There were also operations in the Collie area of Western Australia (employing 765) and in Tasmania (less than 300). Victoria also had a significant brown coal industry providing coal to power stations in the La Trobe Valley.

The upheavals experienced by employees and their families and communities and the producers during the 1920s and 1930s had been massive. The Depression, the northern lockout and other major disputes, the loss of export markets, widespread unemployment, cut-throat competition, the lack of profits and disasters such as Bellbird and Mt Mulligan had left deep wounds. These upheavals and their impact on the workers in the industry and their communities were undoubtedly major factors in the turmoil which was to come during and after World War Two.

Union's big wins on compulsory retirement and miners' pensions

In January 1940 the NSW Government announced the appointment of a Royal Commission to investigate the case for the compulsory retirement of mine workers at age 60 and the payment of a pension for retired workers. The members of the Commission were Justice Ferguson of the Industrial Commission as the chairman, and one representative each of the employers and Miners' Federation. The establishment of the Commission fulfilled an undertaking by the Government as part of the settlement of the major dispute in 1938. While this Royal Commission's brief was only in relation to NSW, the Queensland Government had agreed to consider its findings and so followed the proceedings closely, as did Victoria and Tasmania.

The Commission reported in January 1941 and recommended that mine workers should be required to retire at age 60. It had found that there were around 1700 mine workers in the industry aged over 60, 179 of whom were 70 or more, and that there was even one worker aged 89.[293] The Commission recognised that forcing workers to retire would open up opportunities for younger unemployed mine workers and youths, with thousands of unemployed living in the major coal producing districts. A pension scheme for retired miners was the other key recommendation of the Commission, with the pension for the miner to be set at £2 ($4) per week, and $2 per week for wives and $0.85 per week for each dependent child. The fund to pay pensions would receive contributions from producers, mine workers and the Government.

The NSW Coal and Oil Shale Mine Workers' Pensions Act of 1941 received Assent in October 1941 and incorporated the key recommendations of the Commission, although it contained a lower rate of pension. Miners who now had to retire and miners who had been retired as permanently incapacitated from February 1930 now qualified for a pension. The Queensland Government introduced its legislation in October 1941 and its scheme commenced in January 1942. Tasmania was slower in acting and it took until 1945 for Tasmanian coal miners to qualify for the pension. In Victoria it took even longer for a scheme to be established and involved some bitter industrial disputes pitting the miners against the Victorian Government who owned and operated the mine.[294]

The NSW and Queensland Acts set up what became known as the Miners' Pension Funds which proved to have long lives. The Queensland fund was merged into QCOS, the Queensland Coal and Oil Shale Superannuation fund, in 1990. The NSW fund was merged into the superannuation fund for NSW coal industry, Coalsuper, in 1995. Coalsuper and QCOS merged to form the Auscoal Superannuation Fund in 2005; the Auscoal fund was renamed the Mine Wealth and Wellbeing fund in 2015 and is now simply Mine Super.

The issue of compulsory retirement of mineworkers was generally accepted in the industry for many years. However when mainstream Australian industry began to change and anti-discrimination legislation effectively meant that it was illegal to force workers to retire at a certain age, the writing was on the wall for the coal industry schemes.

Many of the miners who retired in those first years of the operation of the pension schemes had led long and interesting lives, much or all of which was spent in the coal industry. And many of those aged 60 or over were ready to be pensioned off, because of their age, injury or poor health. However some were still very fit and well, despite their mature years. One of the miners forced to retire in 1942 was Alf Eke, aged 83. Alf was from Northumberland in England and went to work in a colliery at the age of 10, spending a 10 hour day underground as a clipper, coupling and uncoupling skips to or from the conveyor. Alf graduated to the status of a fully-fledged coal miner at age 15 and later migrated to Australia. In 1942 he was still working for Hebburn No.1 colliery in the Newcastle area as a deputy on the 'dogwatch' shift, the night shift which ran from around 11pm to around 7am. His job involved touring the mine on his own, examining the workings, inspecting the roof and sides of tunnels, including looking for evidence of flammable gas. He was about 180 cm tall, well built, and with an appearance that suggested a man years younger.[295]

The oldest miner in NSW at the time was James McLuckie, who turned 89 in June 1942. He was borne in Ayrshire in Scotland and had been working constantly in the coal industry in Scotland and Australia since he was 9 years old. His first job in Scotland was as a trapper, opening and closing doors underground. He came to Australia when he was 20, finding work in AACo's Borehole colliery at Hamilton, where the first miners' lodge was formed in 1857. His last job before retirement was working on the conveyor belt at the Stockton Borehole Collieries' Boolaroo colliery. James said that he was only going to retire because the new law forced him to do so.[296]

With the new schemes in operation, pensions began to be paid quickly. By July 1942, the Queensland fund was already paying pensions to 574 people - 294 of whom were miners, and the balance widows, wives or female dependents and children.[297] In NSW the pension fund was paying pensions to around 1500 retired mine workers within the first year of its commencement, and by the early 1950s that number had grown to more than 8550.[298] With the Commonwealth Government pension in the 1940s only available to workers when they reached the age of 65, the coal industry funds in NSW and Queensland were a godsend for the miners and their families who became pensioners at age 60, or who were aged under 65, and who now had retired. The pension schemes also provided for widows and dependent children under 16 years of age to have an ongoing entitlement to a small fortnightly pension, with the children's pension ceasing at age 16. "The peace of mind provided by their pension entitlement for widows and children was profound. That peace of mind…came during an era when the death or crippling injury of a breadwinner often meant near-immediate destitution for surviving family members. With no money to pay weekly rent, evictions were common in mining communities. Meagre household goods were thrown into the streets. Suicides, though seldom spoken of, destroyed the remnants of already broken families…The miners' pension, when it came, was often, quite literally, a life-saver."[299]

Menzies meets the striking miners

In 1938, the industry's total production was around 11.7 million tons. The following year production rose by around 2 million tons, but in 1940 it fell back to around 1938 levels, mainly as a result of a 10 week national strike in March, April and May. Coming at a time when demand for coal for the war effort was critical, the 1940 strike had major repercussions throughout the country. This strike cost the industry nearly a million tonnes, and saw stocks held by the main users including the railways severely depleted. Electricity and gas restrictions applied during the strike, steel production was affected, and according to the Sydney Morning Herald, around 20,000 workers in other industries were thrown out of work during the strike.[300]

The strike had its origins in the coal strike of September 1938 which led to the historic decision in 1939 by the Arbitration Court's Judge Drake-Brockman to grant claims by the miners, including a 40 hour working week for the whole industry. The decision was an interim one and its application was postponed in NSW. In October 1939 the Full Court, by a majority (with Drake-Brockman dissenting), decided that the 40 hour week which the miners were seeking should only apply to underground workers and some surface workers. The decision and the subsequent Full Court hearing also involved the tricky question of by how much wage rates should be adjusted to compensate mine workers for the shorter working week. In 1940 the unions were demanding that the 40 hour week should apply to all workers in the coal industry, that there should be no wage reductions as a result of the shorter working week, and that penalty clauses in awards relating to annual holidays should be removed.

In the middle of the strike in April 1940, Prime Minister Menzies travelled to Kurri Kurri in the heart of the northern coalfields to address the local community.[301] The local picture theatre was the venue, but the miners had earlier voted to boycott the meeting. Menzies spoke to a small crowd at the picture theatre, and also met a delegation of local community leaders. He then proceeded to the local sports ground where the miners were holding a rally. At the sports ground he waited until the leaders of the union had finished addressing the audience of 3,000. Menzies then tried to convince the miners that the issues involved in the strike were minor in the context of the war effort and the impact the strike was having. However the miners obviously did not buy Menzies' argument and a motion was carried unanimously that the meeting "repudiates the threats of the Federal and State Governments and the

Prime Minister, in his endeavours to break the loyalty of the workers to their elected leaders, and are determined to carry on this fight to a successful conclusion."

Statements by the miners' leaders and Menzies himself were interesting in shining a light on their attitudes to the arbitration system, the war, and in the case of the Federation leaders, to their dedication to communism. Charles Orr, the Federation's national secretary, proudly stated that "I am a Communist because my life experience teaches me that we can get nowhere under this present system. I shall continue to try to defeat the present system despite any threats by Mr. Menzies." Henry Scanlon, the district vice president, was dismissive of Menzies' appeal to the miners' patriotism: "We have one war now, and it looks as if we'll have another." Thomas (Bondy) Hoare, the Federation's northern district president stated that he did not intend to agree to arbitration of the dispute. Hoare said that the Federation had never been an "arbitration union", but relied on its industrial muscle, and that it would continue to hold the line until it had won this battle.

This stated attitude of opposition to the arbitration system and preference for direct action of course came after the system itself had granted the miners major gains in wages and conditions, although it failed to give them all they were seeking. However the anti-arbitration stance should not have been a surprise as the election of communists in 1934 to the leadership of the Federation largely centred on their campaign for direct action and repudiation of the stance of the old leadership.

Menzies was determined to keep the miners and the dispute within the arbitration system, telling the miners that they could require the Arbitration Court at any time to call a compulsory conference, with the Court able to issue orders if necessary. At a meeting with the local delegation earlier in the day, Menzies rejected a suggestion that there should be a secret ballot among the miners. He said there would be no point, and that the Government was not prepared to legalise a strike which was called to oppose an award of the Court just because of a secret ballot. He was also doubtful about the value of a compulsory conference in this dispute, saying that before the strike had begun, he had attended a conference involving the unions and employers, and said that he had "never presided over a conference in my life that showed fewer symptoms of coming to an agreement about anything. The information that I have had since suggests that a conference of that sort is no more likely to succeed now than then."

In Queensland there had initially been an overwhelming vote against joining the strike, with the vote in the Ipswich and West Moreton district 780 against to 80 in favour, although the workers at Collinsville voted in support of the strike. However following a request from the Miners' Federation in NSW to support their NSW brethren, the Queensland Colliery Employees' Union secretary ordered the Queensland miners out on strike. However after about 8 weeks, with support from Queensland collapsing, a mass meeting of miners on May 8 voted to go back to work the following day.[302]

The strike, which began on 11 March, was settled in May, with the NSW unions agreeing to resume work immediately, and workers were back on the job by 20 May. The return to work was agreed on the basis that the Arbitration Court would settle the outstanding issues. In NSW over 500,000 working days were lost as a result of the strike.[303] The unions also undertook that during the war they would abide by decisions of the Arbitration Court, and that they would only use the industrial machinery provided by law. The union leaders gave an undertaking that that their members would return to work for the duration of the war. These commitments by the Federation's leaders were to be sorely tested in the months and years ahead by the actions of the union members.

First national coal board

Following the strike, and given the importance of coal to the war effort and the normal functioning of the economy, the NSW and Federal governments moved to subject the industry to official control. In May 1940 a Coal Distribution Committee was set up in NSW, "which organised statistics of production and consumption and rationed coal, ensuring supplies to utilities in New South Wales and to the Melbourne

Gas Company."[304] The Federal government introduced regulations on the supply of coal which gave the minister power to require coal owners to supply coal in specific quantities and to specific users, although he was required to have "regard to all the relevant circumstances" when specifying the time within which the coal should be supplied. [305]

The establishment of a powerful new Coal Board and new industrial relations machinery then followed in February 1941. The Coal Board was chaired by Justice Colin Davidson of the Supreme Court of NSW. Its other members were appointed to represent the coal producing companies, the unions, and the major users. Thomas Armstrong, chairman of directors of J & A Brown and Abermain Seaham Collieries Ltd, and Stanley McKensey, superintendent of Hebburn Collieries Ltd, were the producer representatives. Charles Nelson, the Miners' Federation general president, represented the Federation. The NSW Railways and the Commonwealth Department of Supply and Development and the Department of Commerce were also represented. Having chaired a joint Commonwealth and State royal commission into the industry in 1929 and the 1938-39 NSW royal commission which investigated the health and safety of coal mine workers, Davidson was well acquainted with the industry and its problems.

The regulations establishing the Coal Board gave it the power to control the supply, distribution, storage and use of coal, but not the production of coal. In his official war history, *War Economy 1939-1942*, S J Butlin described the Board's functions as effectively "limited to control of coal once it had reached the surface and was available for distribution."[306] However the Coal Board was replaced within a few short months by a new body with much greater powers. In response to the industry's continuing turmoil, the establishment of a Coal Commissioner to replace the Coal Board was announced by Prime Minister Menzies on 7 August 1941.[307] The new body was now to control and direct the production, distribution, supply and consumption of coal. Control of production was now to be in the hands of the government. Menzies said that the new arrangements were intended to rationalise the industry and not nationalise it, with "the whole of the production and distribution controlled in the public interest." Industrial stoppages and disputes had been continuing to plague the industry following the national strike in 1940, with one strike by engine drivers on the South Maitland coalfield still continuing after 10 working days, and with the producers claiming that this strike had already cost around 250,000 tons of coal to be lost.

Norman Mighell, chairman of the Repatriation Commission, was appointed to head the Commission. Charles Nelson and Judge Drake-Brockman were appointed as consultants to the Commission, but following a decision of the Miners' Federation central council which objected to the powers of the new body, Nelson resigned from that role late in August. The powers of the new commissioner were very broad and covered the control of the production, treatment, handling, supply, distribution, storage, marketing and use of coal. R P Jack, senior inspector of collieries in the NSW Department of Mines, was appointed the Commission's chief executive officer. The other senior staff included Stanley McKensey, who was appointed production manager, and H Leman Williams, managing director of Brown's Coal Pty Ltd in Victoria, who was appointed distribution manager.

The Government also established a new industrial body for the industry – the Central Reference Board. The new Board began in February 1941 with Judge Drake- Brockman as its chairman. Drake-Brockman was a judge of the Commonwealth Court of Conciliation and Arbitration but in his new role was able to operate independently of the Arbitration Court.[308] Other members of the Board were appointed to represent the coal employers and the unions. The Central Reference Board's powers related to any dispute, although Local Reference Boards were also set up to settle disputes arising out of awards by the Arbitration Court or the Reference Board, and disputes referred to them by the Board.

Menzies lost support within his Coalition and was forced to resign as Prime Minister on 28 August 1941. The Coalition government continued for a brief time with Country Party leader Arthur Fadden as Prime Minister. But in October 1941, a Labor government took office with John Curtin as the new Prime Minister. Just 4 months later, the new government replaced the Coal Commissioner with a Coal Commission which now included a representative of the miners' union as a member. Mighell continued as the chairman of the new Coal Commission.

Japan now a direct threat to Australia

With the Japanese attack on Pearl Harbor on 7 December 1941, the fall of Singapore to the Japanese on 15 February 1942 and, three days later, the first of over 60 Japanese bombing raids on Darwin and attacks on other Australian targets, the war took on a new complexion for Australia. Australia was now under direct threat and our industrial and transport capacity needed to increase. However the coal industry continued to be a major problem for the country, with regular strikes and high absenteeism having a dramatic impact on coal production. While the level of disputes in 1942 fell compared with earlier years, around 13% of the industry's production was lost because of the combined effects of strikes and absenteeism.[309]

'War with Japan placed the industry in an even more vital position than it had been before. The expanded munitions programme depended ultimately on coal. So, too, did the railways, coastal shipping and many other essential services. War in the Pacific made vital the accumulation of adequate stocks of coal at strategic points throughout the country. No government whatever its source of political power could tolerate, in these circumstances, any avoidable interruption to the working of the mines. In short, a dramatic improvement in industrial relations was expected...'[310]

In February 1942, with 11 mines in NSW idle because of disputes, Coal Commissioner Mighell put the boot into the miners. "With the Japs at our very doors these stoppages seem to show a complete lack of realisation of the dangers confronting us" he said. Mighell said that while he had no authority in industrial matters, which were the responsibility of the Reference Boards, he believed that as these boards had been set up to deal with disputes, the number of stoppages should be minimised. Mighell noted that the Victorian Railways' operations had already been reduced, and that unless there was a considerable improvement, Australia's war effort would be seriously impaired. He objected to the practice of miners striking in order to attend meetings, or because the industrial tribunals, on which the miners were represented, had not yet finalised a decision. He also said that the fault lay with the miners rather than their leaders: "I know the elected leaders of the Miners' Federation are quite honest in their desire to maintain maximum production, but it seems obvious that some mine employees pay as little attention to the appeals of their leaders as they do to the urgent requirements of the country."[311] The inability of the Federation's leaders to control the actions of the union's members would be a continuing theme during and after the war, including with the major inquiry into the industry chaired by Colin Davidson in 1945-46.

Government strengthens controls on the industry

With continuing and severe problems in the industry, the Commonwealth Government introduced regulations in 1942 to allow unions to expel members who refused a union direction to return to work. Colliery owners were also prohibited from closing a mine without the permission of the Coal Commission. Other regulations allowed the Commission to deal with workers who unreasonably refused to work by directing that they be called up for military or labour service[312]. According to Butlin these regulations were a reaction to the uninterrupted progress of the Japanese and he concluded that 'In these grim weeks there was need for absolute discipline to be enforced by the most stringent penalty.'[313]

Curtin's frustration with the wildcat strikes in the industry was clear in his statements to Parliament in 1942. On 8 May when the Coral Sea Battle was underway, he had warned the country about the grave situation facing the country and asked the Australian people "to make a sober and realistic estimate of their duty to the nation" and appealed for the "the maximum support of every man and woman in the Commonwealth".[314] A few days later, with many coal mines idle due to disputes or to workers simply deciding not to work, Curtin went to some length in detailing the situation in the coal industry.[315]   He explained that there had not been a stoppage in Victoria for many months and that Tasmania had not seen a strike, with its miners "more fully employed than ever"; Western Australia was the only State in which miners were working six days a week, with those miners also having worked every public holiday that year. Curtin said that "the problem of lost production is confined to one State". That State was NSW where over half a million tons of coal "which could have been mined in New South Wales during the present calendar year had not been produced." Curtin quoted a number of examples of mines which were idle even though there had been no formal dispute or where the union had not approached management about concerns. These mines included large mines in the Newcastle district such as Hebburn No.2 where the miners "merely assembled at the pit-top and decided not to work"; Aberdare where "the men held a pit-top meeting this morning, and left the mine"; and Stockton Borehole where "the men returned home this morning without approaching the management, and the mine is not working." At Burwood colliery Curtin said that the workers had been on strike, with about three quarters then going back to work underground; the other workers then assembled at the pit top and "although the responsible officers did everything to get them to go to work, they refused to do so." Curtin finished his statement that day by calling on the mine workers to go back to work the next day, but warning them that "if they do not go back, the Government will invoke all its authority to compel them to do so."

Later that month Curtin took a different approach to the industry and convened a conference of representatives of the coal owners and unions. He wanted to see if agreement could be reached on how to maintain the necessary level of coal production. The outcome was the Canberra Code, an agreed set of rules and procedures for preventing and settling disputes. The code involved establishment of union/ management committees at each colliery to discuss disputes. Where agreement was not possible, disputes were referred to a higher level union/ management structure, and then if necessary, to a Reference Board, with decisions of the board to be binding, and no strikes were to occur until the laid down procedures had been followed. Davidson noted that "The code was a sensible and flexible arrangement that might have been effective in its purpose if organizational deficiencies had been a major cause of poor industrial relations in the industry. But the causes of conflict were more deep-seated."[316]

The Government's objective was for the industry to produce over 15 million tons a year, and in July 1942 it tightened regulations to make it an offence for one miner to incite another miner to stop work. This led to some prosecutions and seemed to reduce the level of stoppages in the second half of the year. However the situation became so serious that the Commonwealth Government was forced to introduce coal rationing in October 1942. Conditions did not improve in 1943 with the country again facing a severe shortage of coal by the winter months.

Production had almost reached the target level in 1942, but 1943 saw a reduction of about 800,000 tons, or over 5%. Demand for coal in the year to mid-1943 had increased by around a million tons to 15.5 million tons, with the demand from the railways and the troop movements the major factor. Butlin concluded that "the coal shortage was due mostly to the failure of the industry to operate at maximum feasible capacity" and that 'Most of the shortfall in (1943, 1944 and 1945) was due to strikes and avoidable absenteeism." Butlin also noted that fatigue was a problem in the coal industry and was aggravated by poor working conditions, and that the increase in the average age of miners also had a negative impact on the efficiency of the industry. Younger men were enlisting in the armed forces in 1941 and 1942 and were being replaced by older men, many of whom had been unable to obtain work during the Depression. He also pointed to equipment deteriorating during the war, although this was not unique to the coal industry.[317]

Curtin rejects Federation proposals

In 1943, the Miners' Federation submitted a package of "practical mining" proposals to the Coal Commission and the Federal Government covering a range of its concerns. Curtin made the Government's response to the Federation public, no doubt in the hope that it would expose the union to more public pressure to ensure that its members concentrated on producing the coal vital for the war effort. The Federation's proposals included a demand for an increase in the number of skips in mines, a requirement to make horse wheeling compulsory in Queensland mines, machinery to be provided for the Blair Athol and Commonwealth No.2 open cut mines, pit committees to be made compulsory, and power boring machines to be provided for miners. These and the other proposals were rejected by the Commonwealth Government, with the Government's reasons illustrating the difficulties in achieving change in the industry. In relation to the number of skips, the Government argued that in most mines some workers did not fill their darg (production or work quota) each day for a variety of reasons including mechanical breakdowns, but that the major problem was that some wheelers and clippers hurried to get skips to their colleagues who load the skips so that the workers loading the skips could fill their darg easily, allowing the whole team to leave the mine early. This created dissatisfaction amongst other mine workers who left at the same time, whereas if they had waited, there would have been ample skips. Also it was pointed out that most mines employed a Federation member to control the skips, but who was not able to control the men working under him, with those workers doing as they liked. "As a rule, any vigorous attempt to enforce discipline leads to a stoppage of work."[318]

Power boring machines were one of the Joint Coal Board's early priorities after it was established in the late 1940s, however during the war it was not easy to purchase mining machinery, with the Government saying mechanisation was a slow and complicated process and to obtain the advantages it offered would require the complete cooperation of mine workers. Given the resistance to mechanisation in the past, the Federation's proposal for power boring machinery was difficult to understand. On the question of pit committees, the Government noted that it did not believe that this proposal would reduce stoppages when lodge members were in the habit of refusing to pay regard to decisions reached by such committees.

While the Government did accept not the Federation's proposals, or pointed out that in some cases what was proposed had already been addressed, it did make important changes that year to the industry's industrial relations machinery, largely in response to the Federation's concerns about the decisions of the Central Reference Board. A Central Coal Authority was established, with this new body having responsibility solely for matters involving the Miners' Federation. Albert Willis was appointed to chair the new Authority, and was also appointed as a conciliation commissioner for the duration of the war. The other members appointed to the board were R W Davie, the secretary of the Northern Colliery Proprietors Association, William McNally, representing the combined western and southern colliery proprietors, and the general president and general secretary of the Federation (those positions were vacant at that time due to union elections). Drake-Brockman continued to operate as the Central Reference Board, but his responsibilities were now restricted to matters involving the other unions. The Local Reference Boards gained in power as they were no longer subject to the authority of the central body.[319]

Albert Willis was the first general secretary of the Miners' Federation when it was established in 1916 as the Australian Coal and Oil Shale Employees' Federation which brought together the NSW Federation and Queensland Coal Employees' Union members. Willis was president of the NSW branch of the Australian Labor Party from 1923 to 1925; he resigned from his Federation role in 1925 when he was appointed to the NSW Legislative Council; he also represented NSW as its Agent General in London in 1931-1932. Before his new appointment Willis had been the liaison officer between the coal industry and the Manpower Directorate.

The Curtin Government also restructured the Commonwealth Coal Commission in early 1943, with a single Coal Commissioner, reverting to the structure of 1941. Mighell was appointed the Commissioner in March 1943, with the additional power to take over the operational control of any mine and pay compensation to the owners. Fortunately, despite the changes made to the Coal Commission since it was established in August 1941, continuity had been maintained through Mighell's ongoing role as the head of the organisation.

Calls for the nationalisation of the coal industry were also prominent in 1943. Eddie Ward, Curtin's Minister for Labour and National Service, told a meeting in Perth that he had recommended to the government that it should take over the industry, and that once the Labor Party got control of both Houses of Parliament it should also nationalise banking and insurance.[320] Curtin and Ward were political enemies, and Ward's call for nationalisation of the coal industry was rejected by the Government. The Miners' Federation also demanded that the Government nationalise the industry or make it subject to full Government control, with a Federal Minister for Mines having responsibility for all coal production and distribution and industrial relations matters. The Government's response was that it already had full power under the wartime National Security Act to ensure that sufficient coal was produced and to allocate the coal as required. Nationalisation would mean compensating the mine owners and this would simply increase "the colossal war budget."

By the winter of 1944 the situation was worse than in 1943, with production lower than in 1943. Since becoming Prime Minister, Curtin had been personally involved in significant negotiations with the Miners' Federation, but was clearly extremely frustrated. In a statement in the House of Representatives on 8 March 1944 his frustration was clear: 'So much has been said on this subject that there is not much more that I can say. The facts appear to me to be these: the Miners not only refuse to respect the wishes and policy of the Government, but they also refuse to respect the advice of their own leaders. A mine is idle today because one locomotive driver in the mine said that another locomotive driver was endeavouring yesterday to put something over him. Circumstances associated with stoppages of that kind cause me great anxiety. They do not arise out of the normal disputes unions have with employers over claims for improved conditions or for higher pay. Frankly, I do not know the basic cause which produces this state of mind."[321]

But Curtin's frustration did nothing to overcome the seemingly fruitless search for ways to induce the industry to concentrate on its major task of producing sufficient coal for the war effort. "Whichever way the

Government turned, whether to the imposition of severe penalties, or conciliation and concessions, to alterations of the industrial machinery, or to exhortation and the offer of inducement, the end result was still a steady rise in industrial turbulence and decline in coal production.'[22] World War Two ended with the surrender of Nazi Germany in May 1945 and the surrender of Japan on 2 September 1945. Australia now faced the task of getting its people and its economy back to normal and beginning the process of planning for post war growth and removal of wartime regulatory controls. But the Australian coal industry remained a basket case, unable to meet the needs of the nation.

Coal cutting machine in Burwood Colliery 1946. NAA A1200 L6941

Miner in confined space late 1800s. (Courtesy Mine Super)

5

The industry at the end of World War Two

The mines and companies

At the end of the war, the NSW industry was a mixed picture in terms of its structure and the degree of mechanisation in mines. The industry was dominated by a small number of major companies including the BHP/ AI&S group, JABAS, Caledonian and Hebburn. A number of mines were fully mechanised and some others had progressed some way along the path to full mechanisation, but there were many mines which were substantially unmechanised.

In 1945 NSW mines produced only just over 10 million tons of coal. The NSW industry's potential production was much higher, as evidenced by the fact that in 1945 around 2.3 million tons were lost to industrial disputes, and a further 1.2 million tons to absenteeism. Total losses from disputes and absenteeism therefore were 35% of actual production. For Australia as a whole (with Queensland accounting for most of the production outside NSW), the comparable losses were around 17% of production.[323] NSW production picked up in 1946 to 11.2 million tons, but it would not be until 1952 when NSW would be producing sufficient coal to avoid shortages, although this was just in time to see a surplus emerge as market conditions changed.

Employment in 1945 in the 139 mines in NSW totalled almost 17.500, with almost 13,000 of these workers employed underground, the underground workforce including just over 400 males aged under 18. The workforce in both NSW and Queensland was 100% male, the official statistics in both states referring only to the number of males employed in the industry. Presumably the only females to be found in the industry would have been employed in company offices away from the minesites in centres such as Newcastle, Sydney and Brisbane.

The NSW coal industry at the end of the war was selling around 2.8 million tons a year to major customers in other States, with most of this coal shipped from the port of Newcastle. Exports to other countries, which had peaked at around 3 million tons in 1912 and which had fallen away in the 1920s, averaged only around 300,000 tons between 1944 and 1946.[324] It would not be until the late 1950s that exports would start to regain an important position as the Japanese steel industry began to take a close interest in Australia as a coal supplier. Queensland exports at the end of the war were negligible; in 1949, the first year for which the Queensland Coal Board produced statistics, exports to other states were zero and exports to other countries less than 4,000 tons. Queensland did find markets in other states in the next few years, but significant exports to other countries would not occur until Thiess Bros began supplying coal to Japan in 1961-62.

Queensland employed around 3200 workers at the end of the war, with the Ipswich district accounting for just over half of the total. The Bowen district employed 500 (at the Bowen State and Bowen Consolidated mines), the Maryborough district 315, the Rockhampton district 250 (including the Styx State mine), the Darling Downs district 235, the Clermont district 147 (the Blair Athol open cuts), and the Chillagoe district 88 (the Mount Mulligan State and King Cole mines). There were 85 collieries recorded as producing coal in 1946, 56 of these in the Ipswich district. The state produced around 1.5 million tons in 1946, all but just over 100,000 tons from underground mines. In 1946 there was only one underground mine producing over 100,000 tons per year (Collinsville State – 173,000 tons), and only three other mines producing over 50,000 tons (Bowen Consolidated - 60,000, Rhondda - 65,000, and Rylance No.3 - 52,000); the next largest was the Styx State mine (42,000 tons).[325]   Queensland was dominated by family owned mines, with owners often also working in their own mines. As we will detail later, most mines were primitive, with little or no mechanisation, and many would never be suitable for full mechanisation because of their small size, poor design or other factors.

Housing and mining communities

In January 1945 the Commonwealth Government decided the time had come for a major investigation into the industry and it appointed a Board of Inquiry to look at a wide range of issues including absenteeism, stoppages, the system of industrial relations and health and safety of workers. The original members were the chairman, Justice Colin Davidson, Thomas Armstrong, representing the colliery proprietors, and Idris Williams, president of the Victorian branch of the Miners' Federation, representing the employees. Williams resigned due to illness in January 1946, but the Federation declined to replace him. Armstrong resigned due to illness in June 1945 and was replaced by David Robertson, managing director of Wallerawang Collieries. The Government decided to appoint Davidson as sole commissioner in March 1946, and he presented his report to the Government later that month.

Davidson's findings and recommendations are considered in the next chapter, but to set the scene for how the industry was placed as it entered the new post war era, let's look at what life was like for miners and mining communities at that time. At the end of the war, miners and their families typically lived close to the mine in which the father, and often the son, worked. Commuting to work in the pre and post war years was not the norm. "As most mines were small and dependent on coal orders for survival, most had low stocks and work continuity depended on demand. The colliery whistle was most meaningful to the miners within hearing distance. When it blew in the late afternoon it signalled that production was needed on the next working day and work was available, and the miners were expected to arrive. The whistle was also used to signal serious accidents which also drew relatives to the pit …"[26] Mine workers were employed on a daily basis, but this would change in due course when the award was amended to provide for weekly employment.

Housing conditions in mining communities varied considerably, from disgraceful to satisfactory. Some of Davidson's observations about housing give an insight into those varying conditions: "At Cessnock … a group of miners' residences is noticeable on the outskirts in the 'depression era', as it is called, that are totally below a proper standard and should be replaced. During the depression period … mine workers were forced by circumstances to live in these 'shacks', as it was believed only temporarily, until trade recovered. But, as usually happens, the temporary structures became permanent …. although latterly rendered more habitable when reticulated water and electricity were made available."[327]

"… the living conditions at Catherine Hill Bay … are scandalously bad. At this colliery some of the mine workers live in hovels which are disgraceful, but the company which owns the property is not entirely to blame as it appears to be unable to reach any agreement with its employees as to the provision of suitable homes acceptable to them."

'The Southern and Western Districts of New South Wales are, as a whole, not open to such strong comment as regards the housing of the miners. …(However) In the Western District … there are also outlying mining settlements where the living conditions are very bad and should receive attention."

Davidson also laid the blame for poor housing conditions at the feet of the producers, including the NSW Government: "…it is popular practice to charge the owners of collieries in New South Wales with flagrant disregard of the interests and well-being of the employees by failing to expend money on housing schemes and the improvement of their living amenities…. Colliery proprietors have undoubtedly been short-sighted and have brought upon themselves much deserved criticism. But they are not alone to be blamed. When Governments have been the owners of collieries, they have done no more than private owners in this respect, and sometimes much less."

But Davidson also commended the actions of mine owners in two localities: "At Muswellbrook, the company operating two old mines has assisted its workers financially in purchasing their own homes, and at Kandos … the company has been the means of its employees enjoying amenities that are obviously highly appreciated … at the latter colliery, which is as fully mechanised as possible, the annual output per employee for 1944 was 2,100 tons of coal as compared with 700 for the whole State." He also had good things to say about the Miners' Federation in Victoria: "At Wonthaggi, the effects of the District Branch of the

Miners' Federation have been most effective in improving the living conditions of its members. By means of contributions made fortnightly by the miners themselves, free medical and dental attention has been established, whilst a moving picture theatre, a workmans club, meeting halls and other amenities have been provided for their benefit. "

In Queensland Davidson found both good and bad. Housing conditions at Ogmore and Howard were good according to Davidson; in Rosewood they were reasonable. Davidson found that the best housing and living conditions in Australia for mine workers were in Ipswich. But in Collinsville and Blair Athol, he found that "living conditions and amenities had been sadly neglected."[328]

Miners' health was poor

The Joint Coal Board, which began operations in 1947, put the need to increase coal production at the top of its priorities. However it was also vitally interested in the health of mine workers and the abysmal standard of amenities at most mines. One of the most interesting, but alarming, aspects of the JCB's first annual report was the detailed data presented on the standard of amenities for mine workers above and below ground. The JCB surveyed amenities and found that the general standard in NSW was "deplorably low". In mid-1948 there were 152 underground mines and 14 open cut mines operating. Of the underground mines, no mine had a crib room (an underground area for meal breaks) which was up to standard; only three mines had adequate dust suppression systems; only two had adequate sanitation: and only seven had adequate water reticulation. In relation to surface facilities, only three mines had bath and change houses which were of a reasonable standard; and only 14 had adequate sanitation.[329]

Given the poor standard of both housing and mine amenities, it is not surprising that the health of many miners was poor. In 1948 the JCB conducted a sample survey which covered over 1500 mine workers and found that 16% had pneumoconiosis (black lung disease), 16% had recurrent bronchitis or chronic lung disease, and 9% cardiovascular or renal disorders.[330]   These disturbing findings would see major programs introduced to improve health standards.

Davidson also said that the health of mine workers was another area of major concern, particularly in relation to dust disease: "Probably the most important of all problems concerning the health of workers in coal mines underground and in the vicinity of the surface screens relates to the control of dust. Despite the legislation in New South Wales which implemented the recommendations of the Royal Commission in 1938-39 the incidence of disease from the inhalation of the finer particles has become positively alarming."[331] But Davidson added that "Blame must be attached to a world-wide deficiency of knowledge rather than to the management for this unhappy state of affair."[332]

Despite the Royal Commission into mine safety in 1938 and the major amendments to the NSW Coal Mines Regulation Act in 1941, and Queensland's amendments to its Act in 1938, the wartime years saw little real improvement in the industry's safety record. Deaths continued at an alarming level, although there were signs after the war that the trend was improving. During the years 1939 to 1945 there were 143 fatalities in NSW coal mines, an average of over 20 per year, with falls of coal or stone from the roof or sides of underground mines accounting for one in every two of those fatalities. The NSW death rate was around 1.4 deaths per year for every 1,000 mine workers. The trend at the end of the war and just after was better, with an average of around 14 fatalities per year from 1945 to 1948; however 1949 saw the number jump again to 24. Queensland saw 22 fatalities from 1939 to 1945, or around 1.2 deaths per year for every 1,000 mine workers, with falls from roof or sides again the major cause. In 1946 Queensland saw 7 fatalities and 3 in each of the following two years, with no fatalities in 1949 or 1950. Both states clearly still had a long way to go to eliminate fatalities and make their mines safe. New technologies that spread through the industry in the 1950s, 1960s and beyond would play a major role, as would the introduction of modern management systems and regulation which required a risk based approach adapted to the conditions in each mine.

Contract mining system a blight on the industry

Up until mechanisation spread through the industry in the 1950s and 1960s, and new methods of paying mine workers were introduced, the coal industry ran on the basis of the contract system. This system involved the miner (or to use the old term, the hewer), undercutting the coal seam with a pick, and installing short timber poles (props) to support the seam. As mechanisation of the production of coal started to spread in the early 1900s, the undercutting in some mines was done using electrically driven machines or compressed air driven machines. The miner would then drill holes into the coal seam and place explosives in the holes. The explosives were then detonated and the loose coal was loaded manually into skips. Each skip was identified by a small leather tag (called a token) and the miner was paid according to how many tonnes of coal he produced. Wheelers collected the filled skips, and using a horse, hauled the skips to an area called the flat, a marshalling area from where a conveyor would haul the skips to the surface of the mine. Wheelers also brought empty skips back to the miners. Miners were allocated a location in the mine by a ballot, called a cavil, usually held every three months. This ensured that miners shared the more difficult areas in the mine, where their daily production might be affected by poor geology or other factors.

There was strong support for the notion that the contract system caused many problems for the industry. Davidson was clear that abolition of the contract system, along with mechanisation, was key to the industry developing and becoming efficient. The Joint Coal Board would also see the system as a major barrier to improving production and efficiency. In the years after the war, as mines mechanised and the role of the various mine workers changed, the contract system changed to a system where miners' wages were paid on a daily and later a weekly basis, and not on the basis of how much coal each was able to produce. Where a miner previously had to undercut the coal seam with a machine, bore holes into the seam for explosives, detonate the charge, and then load the coal into skips, new machines revolutionalised the process. "As mechanisation proceeded on a day-wage basis, with bonus incentives for production, it steadily eliminated the industrial chaos inherent in use of the hand mining contract system which had become virtually unmanageable due to the multiplicity of customs and practices, national, district and local agreements, written and oral, and the assumed jurisdictions of the various parties in day to day matters."[333]

The management of the industry, particularly industrial relations, was extremely difficult but it would gradually come to be based on sounder principles relating to efficiency and the provision of sensible incentives. However at the end of the war, the Miners' Federation was still strongly opposed to mechanisation on a number of grounds, including safety issues, but particularly because of the threat it posed to jobs. The producers on the other hand, still dependent on government subsidies to operate and fearful that these subsidies would cease at some time, were reluctant to invest or simply did not have access to finance. There was also a Ministerial ban in NSW on the use of machines to extract the valuable coal left in pillars, one major effect of this being that the producers were reluctant to invest in machines which could only be used for part of the mining operation. It would be a difficult transition in the years ahead.

In Queensland in general the miners, who actually mined the coal, and their assistants, who were in many cases responsible for the first stage of its transport, were paid on a piece-work basis, while the majority of other workers were paid on a day wage basis. The piece work system was a contract system, with the workers paid on the basis of their production or quantity of coal moved. The consultants Powell Duffryn, who undertook a major review of the industry in 1948-49, were not as critical of the contract system as Davidson or the JCB, noting that there was a considerable difference of opinion on the relative advantages of the piece rate and day wage systems, particularly in relation to mechanisation in mines. Powell Duffryn said that those in favour of the day wage system included the large producers in NSW and the majority of producers in the USA who pointed to the successful results achieved by crews operating machinery at the coal face and who were paid on a day-wage basis. These producers argued that it was not necessary to have any form of incentive payment in order to achieve good results. Powell Duffryn pointed out that in Australia the use of this type of machinery was very limited and with its operation involving less manual labour, competition among collieries to secure a job as a member of one of these crews was itself a form of incentive. The consultants went on to conclude that there was considerable room for doubt as to whether the same results would continue to be achieved when all collieries were largely mechanised.[334] As we will see however, the post war years would see the elimination of the contract system, the rapid mechanisation of the industry and the also rapid spread of new incentive payment systems. The industry was about to be revolutionised.

A brief profile of some of the larger mines

A book by Elford and McKeown provides an interesting summary of the operations of some of the larger mines which were operating in 1947.[335] The details below of a few of these mines may assist to understand the way in which the industry was operating at that time. The bord and pillar system of underground mining has been mentioned earlier and was the system used in almost all of the mines in Australia in the early post war years. "In this method, the coal in the seam is extracted in two distinct stages. In the first winning, or solid working, 25 to 30 (per cent) of the mineable coal is removed by driving a series of evenly spaced tunnel like openings called bords. Between the bords, solid coal (pillars) remains in position to support the overlying rock. The pillar coal is removed on the second stage. The bords are connected at intervals by openings of similar cross section termed cut-throughs. Dimensions of bords, cut-throughs, and pillars are influenced by factors such as the thickness of over-lying rock and the nature of the floor and roof of the seam." [336]

The typical dimensions of the bords, cut-throughs and pillars were as follows: bords around 5.5 to 9 metres; cut-throughs were around 4.6 to 7.3 metres wide, spaced at intervals of around 36 to 40 metres. Pillar sizes varied significantly, and in Elford and McKeown's book examples given ranged from around 13 metres by 36 metres to 40 metres by 40 metres.

John Darling Colliery[337]

BHP's John Darling colliery was located near Belmont, around 18 kilometres south of Newcastle, on leases totalling just over 4,000 hectares, with half of this area under the ocean. The colliery employed around 513 workers, and 335 of these worked underground. Production was around 1600 to 2000 tons per day from an 8 hour shift. Two seams were worked – the Victoria Tunnel seam, around 200 metres below the surface, and the Borehole seam, around 270 metres down. The Victoria Tunnel workings were fully mechanised, with electric boring machines, coal cutters and loaders operating at the coal face. The coal was cut and loaded by rail mounted machines into skips. The skips, carrying 100 to 250 kilograms of coal, were hauled to a transfer station by electric locomotives and emptied into large bins. The coal was then lifted by a belt conveyor 170 metres long into 3 ton skips which were hoisted to the surface up one of the mine's shafts. Ventilation for the underground workings came from a huge fan driven by a 500 horsepower electric motor. Electric power was generated by the colliery's own power station on the surface which housed 4 generators. The power station normally supplied only John Darling, but could also supply power to the company's Burwood and Lambton collieries in an emergency. The workings on the Borehole seam were only partially mechanised in 1947.

Burwood Colliery[338]

BHP's Burwood was one of the oldest operating collieries on the Newcastle field in the early post war years, with over 1200 kilometres of roadways throughout its underground workings. Burwood was originally developed in the mid-1800s, was later bought by the Scottish Australian Mining company and was purchased by BHP from Scottish Australian in 1932 and closed for a period of time for mechanisation. Located around 9 kilometres south of Newcastle on 1400 hectares of land, Burwood employed around 650 workers in 1947, 500 underground, and produced around 3000 - 3500 tons of coal per day, making it the largest in the country at that time. Operations were on two seams, the Victoria Tunnel and Borehole, with the Victoria Tunnel workings fully mechanised and workings on the Borehole seam still transitioning from hand mining methods.

Coal mining operations at the face on the Victoria Tunnel seam were similar to the John Darling colliery. The loaded skips in Burwood were also hauled by electric locomotives and emptied into bins, but the coal was then moved to the surface by conveyor. Burwood, John Darling and Lambton all produced coal for BHP's Newcastle steelworks. BHP also had an underground classroom in Burwood which was used to teach trainees various aspects of mining.

State Coal Mine[339]

The State Coal Mine, located almost 3 kilometres south of Lithgow on around 5,000 hectares, was another of the State's large collieries. Owned by the NSW Government, the State Coal Mine employed 490 workers and produced around 1650 tons of coal per day. Unlike the BHP mines, the State colliery was largely unmechanised. All coal was hand mined and grunched (blasted out of the coal face with explosives)[340], although power borers were used in some areas of the mine. Loading of the broken coal was also by hand into skips of 1100 to 1400 kg capacity, except for some areas where development was under way and scraper loaders were used. The coal skips were hauled by an endless rope system powered by a 300 hp electric motor. The rope system was also used to move workers underground. Once they had descended down the shaft, the mine workers were transported to and from their work places on trollies operated by the endless rope system, with every fourth trolley fitted with a lever-operated emergency brake which was in the hands of a dedicated operator. The mine also had an electric locomotive which provided another means of transport underground for both workers and coal between the coal faces and the main rope system.

While hand mining was the norm in the State Coal Mine, the productivity of the mine was still reasonable for that time. The miners on contract achieved an average production per shift of 10.4 tons, and were required to bore holes in the seam, place the explosive charges, fire the charges, fill the skips and place timber props as needed. The mine's average output per employee (the total mine workforce) was 3.4 tons per shift.

Aberdare Extended[341]
The Aberdare Extended colliery was located at Bundamba and was one of the largest mines on the Ipswich field. It was owned by Aberdare Collieries Ltd and employed 86 workers, 34 of whom worked underground. Production was around 200 tons per day from two seams – the Aberdare seam with a working thickness of around 4.3 metres and lying around 190 metres below the surface, and the New Found Out seam which overlayed the Aberdare seam. Aberdare Extended was worked from a vertical shaft. It was a mine well known for the fact that the seams dipped steeply, and for the coal being liable to spontaneously ignite, and for the presence of gas in the seams. Mining conditions were therefore pretty difficult. The bord and pillar method was used, with coal hauled in skips on a single track by a rope system which was powered by an electric winch. There was little mechanisation in the mine.

Rhondda[342]
Rhondda colliery near Ipswich was owned by Rhondda Collieries Ltd and was another of the largest collieries in the Queensland employing 100 workers and producing 300 to 400 tons per day. Two seams were worked, around 90 and 100 metres below the surface, and dipping at a 1 in 6 slope. Rhondda was a partially mechanised bord and pillar mine, using coal cutters and power borers. Workers employed on a day wage basis undercut the coal seam and bored holes in the face, ready for the contract miners who charged and fired the holes. The broken coal was then shovelled by hand into skips of 750 kg capacity. Jigs[343] were used for the main haulage task, with horses used for some of the work. The mine had one of the few coal preparation plants in the State which crushed the lumps of coal, and also screened and jigged the coal.[344] These preparation plants were primitive in comparison to the plants we see today at coal mines.

Blair Athol[345]

Blair Athol open cut, around 350 kilometres west of Rockhampton, was operated by Blair Athol Opencut Collieries Ltd. It employed 66 workers and produced 400 to 600 tons per day from the thickest continuous seam in the country of close to 30 metres. Coal was first discovered in 1864 and mining began in the 1890s. When the railway line was extended from Clermont some underground mines opened in the period 1913 to 1918, but open cut mining commenced only in 1937. With an overburden ratio of only 1.5 to 1, Blair Athol was a gold mine waiting to be exploited given the right investment and market conditions, but that would not be for some years. In 1947, the technology of open cut mining was still fairly primitive; no massive draglines existed in our mines in those days. Blair Athol used a variety of machinery including drilling machines, mechanical scoops (to remove overburden), tractors, rooters and bulldozers, with one Marion steam shovel used mainly to remove coal.

Of course conditions in the many small underground mines in NSW and Queensland were very different to those outlined above. With little or no mechanisation, mining was essentially done by hand, with the assistance in some cases of horses. In Queensland in the late 1940s, only one mine had completely done away with hand wheeling (where miners or wheelers hauled the coal skips underground to a certain point in the mine).

Working arrangements[346]

The typical working arrangements for mines in the early post war years included a standard production shift from 7am to 3pm. This was nominally an 8 hour day, with the 8 hours running from the time that the first man on the shift left the surface to the time that the last man returned to the surface. This was also known as 8 hours bank to bank. Workers received a 30 minute meal break and the shift length also included the time taken to travel to and from the coal face or other place of work in the mine. The effective working time per shift for underground workers was therefore only around 6 hours. Some mines also worked an afternoon shift from 3pm to 11pm, but this shift was for maintenance work and not coal production. The night shift, if worked, ran from 11pm to 7am and was also a maintenance shift. Coal production was only possible at that time under the provisions of the coal awards during the main daytime shift. In Queensland the standard five day week had been worked in the coal industry for the previous 10-15 years, with the nominal length of the working day being 8 hours. However at many mines the actual shift length was much shorter, and was as low as 5.5 hours in some mines.[347]

Mine workers received paid holidays on the basis of a day's leave for each 25 shifts worked, up to a maximum of 10 days per year. Workers who were members of the Miners' Federation and who worked more than 225 shifts a year received an additional 5 days' paid leave. Federation members were paid for public holidays; members of the craft unions had an industrial case current in 1947 on their claim for paid public holidays. All employees were entitled to sick leave, with Federation members receiving a maximum of 10 days per year, which could accumulate to 40 days over time if the leave was not all used. Mine workers were issued with a range of protective equipment free of charge as required for the job, including safety helmets, shin guards, goggles, gloves and respirators. Safety boots were provided to workers at cost price.

6

The Joint Coal Board and Queensland Coal Board era commences

Davidson Inquiry finds an industry in crisis

As already mentioned, Colin Davidson handed down his report in March 1946. He described an industry in turmoil and he warned that the industry in NSW "… is drifting towards disaster, and unless some greater degree of cooperation is procured, there may well be a repetition of the depression within the ensuing ten years."[848] He was scathing about the industry, but praised the work of the Coal Commissioner, Norman Mighell: "The policy and administration of the Coal Commissioner were of unquestionable benefit to the Industry and to the war effort."[849]

Davidson found deep-seated problems including a widespread feeling of antagonism on the part of many mine workers towards the management and colliery proprietors; lack of cooperation in the great majority of mines between management and employees; constant industrial troubles in the form of incessant pit-top meetings, absenteeism, stoppages of production and strikes; extraordinary apathy in the most stable sections of workers towards these hindrances to their earning capacity; almost complete lack of discipline; an excessive accident rate, with lack of discipline being a contributing factor; inefficient and archaic methods of mining in the majority of mines due largely to inadequate mechanisation; mounting costs and declining output; lack of stability to guarantee future prosperity; inability of the owners to obtain a reasonable profit; inflated costs of production due to expensive plant being out of operation for the greater part of the year; and a feeling of frustration amongst managers who were prevented by industrial disruption from carrying out their duties.

One particular statistic – the proportion of coal production lost because of strikes and absenteeism – highlighted how far below its potential the industry was operating. Davidson's report showed that in NSW in 1942 the industry lost an estimated 15% of its production to strikes and absenteeism, rising to 35% in 1945. In other States the situation was nowhere near as dramatic, but still serious, with 6.7% lost in 1942, rising to over 17% in 1945.

% coal production lost to disputes and absenteeism

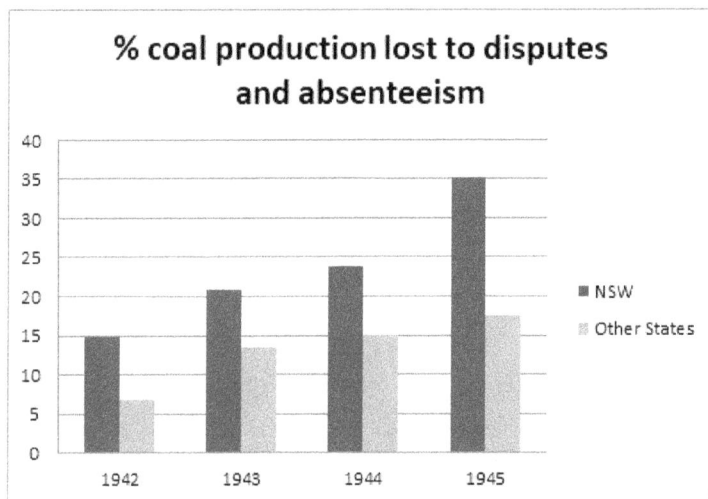

Source: Davidson Inquiry 1946 page following p. 32

Davidson saw a lack of discipline among the mine workers and poor leadership of the Miners' Federation as two of the fundamental problems of the industry. Some of his conclusions were as follows[350]:

"Especially in the Northern and southern Districts of New South Wales, discipline is almost non-existent amongst mineworkers who are members of the Miners' Federation and are within its sphere of influence....Discipline is observed by mine workers who are not members of the Miners' Federation or are remote from its influence and generally, also, by mineworkers in mines that are free from the system of payment on contract rates...Lack of discipline is mainly due to – (a) weak and divided leadership in the Miners' Federation;(b) political antagonism between members who are Communists and those who are opposed to the doctrines and activities of the Communists;(c) political intrigue and ultimate abolition of the Compulsory Arbitration system… (d) the success achieved by nearly all strikes in gaining some concessions;(e) inability or reluctance on the part of the Government to enforce the law against large numbers of individual strikers or absentees;(f) appeasement on the part of the Government in yielding to improper demands under threats of disruption of the Industry; for example in removing judicial officers at the behest of unions which refuse to accept decisions that are adverse…(g) vigorous opposition by the Federation to the dismissal of an employee by the management of mines for any reason whatsoever."

"Practically none of the strikes during the war have been due to impropriety of any kind on the part of the owners or management… Most of the strikes have not been based on genuine grievance, whilst many have not been due to disputes with the owners of management."

"Attempts by the Miners' Federation to impose discipline on its members were substantially a failure, although in a few instances they proved to be effective."

"The system of payment on contract rates is the source and origin of innumerable undesirable customs and practices leading to disputes. "

Davidson was also uncomplimentary about the Commonwealth Government, implying it was weak in its response: "Prosecutions under the National Security Regulations and the Coal Production (War-time) Act, for absenteeism and other offences, achieved a considerable amount of success and might have had more force if pursued inexorably. "

The mine owners were also not spared, with Davidson critical of the state of their employees' housing in many areas and of the amenities underground and on the surface. He said that improving the surroundings and both surface and underground equipment of the pits would 'provide more comfort and … remove causes of irritation …"

A central recommendation of Davidson's report was the creation of a Commonwealth statutory authority with wide ranging powers to collect and publish detailed statistics, work with insurers to reduce the costs of workers compensation and prevent abuses, and appoint expert staff to advise on areas such as safety, coal recovery and mechanisation. The authority would also have emergency powers to arrange for the distribution of coal locally and for export. Another key recommendation was for the new authority to be able to pay subsidies to mines which agreed to come under its jurisdiction. The reason for this was that under the National Security Act during the war, to keep prices at acceptable levels, subsidies had been paid to colliery proprietors, and Davidson saw possible collapse of the industry if this system ceased once the Act no longer applied in peace time: "More than one important colliery would be forced into liquidation, causing serious loss of employment. Large and immediate increases in prices would also be essential…"[351]

Two of the most critical issues to resolve to allow the industry to grow and prosper according to Davidson were the contract system of wages and mechanisation. He said that "The system of payment on contract rates is the source and origin of innumerable undesirable customs and practices leading to disputes." Mechanisation was seen as critical to the future of the industry: "Mechanisation of the mines to the fullest extent practicable would eliminate much of the trouble in the Industry."[352]

Davidson warned against a board for the authority which comprised representatives of mine owners and workers. He said that that would be fatal and noted insightfully: "As always happens in such cases, the chairman, instead of having the assistance and cooperation of independent minds, would spend most of his time acting as a referee between his colleagues, who would probably hold preconceived and mostly irreconcilable views on almost every relevant subject." Instead he said that "The ideal Authority… would … consist of one man with legal training, who would readily acquire the knowledge of the industry required to give balanced decision on various issues."

With the Miners' Federation pushing for nationalisation of the industry, Davidson made it clear he did not see this as the industry's salvation. He noted that nationalisation of the industry had become "the most insistent demand of the unions, supported probably by a majority of the mine workers" and added that in Britain the Parliament had been "induced to accept the responsibility of introducing legislation to acquire all the coal mines upon payment of just compensation. And this decision had been hailed as the forerunner of salvation for the Industry …"[353] The British legislation to nationalise the industry and to place the industry under the control of the new National Coal Board would be passed just a few months later.

Mixed reactions to the Davidson report

Although multiple copies were not initially available to the industry, the key findings of Davidson's report were public knowledge by the end of March 1946; the battle over the report then began. The employers welcomed aspects of the report, but were nervous about the extent of government control. The Miners' Federation was bitterly opposed to Davidson's report; the Federation had initially cooperated with the Inquiry, with its Victorian President Idris Williams one of the members, but once Williams had retired from the Inquiry, the Federation declined to replace him and withdrew from involvement in the Inquiry's work. Following the official release of the report, the Federation's Central Council rejected it, saying it was inspired by "anti working class bias" and that it was a complete contradiction of Davidson's previous findings and of the evidence to the Inquiry.[354]

According to Edgar Ross, one of the major bones of contention between Davidson and the Federation was his refusal to allow the Federation to submit evidence to the Inquiry which supported nationalisation of the industry.[355] The Federation had continued to demand that governments nationalise the industry, with its general president Harold Wells for example telling the union's national convention in August 1945 that this was an urgent necessity.[356] In 1945, Wells had also proposed to the first full sitting day of the Inquiry that it should recommend to the Federal Government that the industry be nationalised.

But with nationalisation clearly not in prospect, in early 1946 the Federation came out in support of the Baddeley Plan. Jack Baddeley, the NSW Mines Minister (and the first national president of the Coal and Oil Shale Employees' Federation in 1916) had been responsible for drafting the NSW Coal Industry Bill in 1930 which failed to receive Assent after the dismissal of the Lang Labor Government. That Bill would have created a statutory body to control the NSW coal industry. From the Federation's viewpoint, State control of the industry was the next best option to nationalisation. The 1930 Bill was re-drafted in 1945-1946 and, if it had become law, would have provided for a new organisation appointed by the State Government, with wide-ranging powers to control almost all aspects of the industry.[357]

The Commonwealth Government now had to find a way forward. In late May 1946 Prime Minister Chifley met with employer and Federation representatives to discuss the way forward. At the meeting with the employers on 28 May, the employers gave in-principle support to continuing Commonwealth control of the industry, particularly control of marketing and distribution. While not expressing complete agreement with the Davidson report, they saw it as the only proposal on the table offering a positive way forward for the industry.[358] The employers saw merit in giving a Commonwealth coal board the power to pay a bounty (effectively a subsidy) to those producers who agreed to come under the jurisdiction of the board, a not surprising attitude given the woeful profitability of the industry over many years, and the fact that producers had been subsidised during the war, with the subsidies continuing immediately after the war. Chifley met the Federation leaders the next day, with the Federation reported as having tried to convince the Government to cede its powers over the coal industry (which still existed under wartime control legislation) to the States. Should the Commonwealth fail to vacate the field, the Federation suggested that it could retain control over marketing and distribution.[359]

Several days later, the Commonwealth announced that it had made a number of decisions in relation to the coal industry, although its formal stance on the Davidson report was not one of the decisions. The Commonwealth would give the Council for Scientific and Industrial Research (the forerunner of the CSIRO) the power to set up research facilities to investigate the prevention of dust in mines, including the power to obtain expert advice from overseas. A committee to come up with recommendations on miners' amenities would be set up, with representatives from the States, the producer associations and the Miners' Federation. Perhaps the most important decision in relation to the strategic direction of the industry in years to come was the decision that, to improve production, immediate consideration should be given to the development of mechanisation in the industry, wherever this was practical.[360]

However in relation to mechanisation, there was little that the Commonwealth could do at that time to stimulate development of the industry, particularly with the NSW Government having incorporated into the Coal Mines Regulation Act in 1941 an effective ban on the use of mechanical equipment to remove coal pillars. Nevertheless, the clear statement from Chifley demonstrated that the Government supported mechanisation as the way forward, and this policy would be adopted as a key plank of its policies by the Joint Coal Board when it began to operate the following year. The Queensland Coal Board would also have mechanisation as a key strategy, but the process would be more drawn out in Queensland due to the structure of the industry.

Chifley Government rejects the Davidson model

Shortly after Davidson had presented his report to the Commonwealth, the Government gave a senior committee of bureaucrats the task of assessing its recommendations. The committee included representatives from key Commonwealth departments and was chaired by H C Coombes, the Director General of the Department of Post War Reconstruction.[361] The committee rejected Davidson's proposed scheme on the basis that a Commonwealth authority would not have the necessary constitutional powers, and also that its success would rely too heavily on the cooperation of the coal producers.

Having previously rejected the Federation's call for nationalisation of the industry, the Commonwealth Government also rejected the option of a NSW statutory authority as proposed under the Baddeley plan. It believed that Commonwealth involvement in the industry was necessary to force the industry to rationalise and modernise. Vacating the field was not an option for the Commonwealth, deciding instead to accept the Coombes' committee recommendations for an authority which would be subject to laws passed by both the Commonwealth and NSW Parliaments. A separate industrial tribunal as recommended by the committee was also accepted by the Commonwealth as the way forward on industrial relations.

The Commonwealth tabled the Coal Industry Bill in the House of Representatives in July 1946. The Minister for Post War Reconstruction, John Dedman, led the debate on the Bill, announcing in his second reading speech the broad details of the legislation which, with parallel legislation in the NSW Parliament, would establish the Joint Coal Board and the Coal Industry Tribunal. The objective of the Bill was "… to provide means for securing and maintaining adequate supplies of coal throughout Australia and for providing for the regulation and improvement of the coal industry in the State of New South Wales; and for purposes connected therewith." Prime Minister Chifley, reflecting the work of the inter-departmental committee, explained that the Government had decided "on the legal advice available to it, that constitutionally the proposals of Mr Justice Davidson revealed so many weaknesses that without complete cooperation, particularly on the part of the owners, they would quite easily become unworkable, because they were largely founded on the principle that cooperation could be obtained."[362]

While the Commonwealth and NSW Governments rejected key elements of Davidson's report, on the question of the make-up of the board of directors of the new organisation they supported Davidson's warning that the new directors (or Members as the Act called them) should be independent of the coal industry. In the debate on the Bill, Chifley told Parliament: "The Leader of the Opposition also referred to the authority proposed to be set up under this bill. I have discussed this matter with the Premier of New South Wales, and I can say that it is the wish of his Government, as well as of the Commonwealth Government, that the proposed board shall consist of three of the best men obtainable, and that they shall be entirely independent of any association of miners or mine-owners."

Joint Coal Board's strong powers

The JCB was given extraordinarily wide powers. While Dedman pointed out that the legislation did not extend to "industrial conscription" he made it clear that the powers vested in the new JCB would allow it to "deal with every phase of the coal industry".[363] These included powers over pricing, distribution, rationing (if necessary), the opening and closing of mines, methods of mining, mechanisation, the health safety and welfare of miners, employment, workers compensation insurance, training, industry statistics and research. Also included were powers to permit the Board to assume control of the management of any coal mine, and to acquire any mine and operate any mine it acquired. Dedman somewhat played down the powers to run and acquire mines: "On the other hand, in some instances, cooperation may not be forthcoming from the owners and managers, and the board may need to assume control of or acquire and operate a mine." However the Board emerged in the next few years as a major mine operator and owner in its own right and did take over operation of some private coal mines. The mining union's demand for nationalisation of the industry was not achieved, but in practical terms, the outcome was that the JCB's operations over the next decade would be equivalent to partial nationalisation.

In the NSW Parliament, Premier William McKell gave a long speech in August 1946 on the Coal Industry Bill, emphasising the nation's dependence on coal and the important position which coal occupied in the national economy, a situation which made "the consideration of some stabilisation of the coal industry of tremendous importance to Australia."[364] McKell also quoted some statistics on coal production and demand which highlighted the challenge for the industry to lift its performance: the demand for coal in 1946-47 was 14.5 million tons, and the expected demand in 1949-50 was 16.5 million tons; however the average output from all Australian coal mines from 1939 to 1944 was less than 13.5 million tons.[365] McKell said that the industry had been "tragically neglected" and that the equipment in the industry was in most cases entirely obsolete, with drab old buildings, workshops in a neglected condition, workers unduly exposed to weather conditions, poor ventilation systems and throughout the industry "an absence of all those amenities that we to-day think of in terms of necessities in association with any great industry."[366] In response to interjections from the Opposition that the Davidson report had been ignored, McKell said that while Davidson did not appear to have considered the option of a statutory authority which combined the powers of both the NSW and Federal Governments, many of the recommendations put forward by Davidson had in fact been incorporated into the legislation.[367]

The other key element of the Coal Industry Acts was the establishment of the Coal Industry Tribunal, which was given responsibility for industrial relations matters affecting the Miners' Federation, including interstate disputes. The Acts also provided for local coal authorities to deal with purely local disputes and issues. The CIT's powers grew over the years to cover all industrial relations issues in the coal industry in NSW and Queensland.

Between the finalisation of Davidson's report and the passage of the Coal Industry Bills by both Parliaments, there was hectic lobbying by the unions and employers and long debates in the Parliaments on key provisions of the legislation. At the end of the process, when the Bills were passed and became law, the Miners' Federation, despite its campaign for nationalisation or the alternative of the Baddeley Plan, accepted the new Acts and their extensive powers. Speaking on behalf of all the unions, Federation president Harold Wells said that the legislation was acceptable to the mining unions, and if the new arrangements were administered well, they would put the industry back on its feet and provide sufficient and regular coal production.[368]

The coal producers, who had been positive about the Davidson model, were less than enthusiastic about the new Joint Coal Board and Coal Industry Tribunal. When the Bill was tabled in June, the producers condemned it as ignoring the recommendations put forward by Davidson, and in fact perpetuating the controls which it argued were largely responsible for the industry's poor state. R W Davie, secretary of the NSW Combined Colliery Proprietors' Association, said that his members wanted amendments to the Bill to provide for industrial laws to be administered by the Arbitration Court, to remove all restrictions on mechanisation, and to give greater control to colliery managers.[369]

The producers were also concerned that the new board's powers would be so wide that it could confiscate assets without adequate compensation. The Southern Colliery Proprietors' Association secretary, William McNally, said that "If this piece of legislation goes through, it means that any shareholder in any company may lose the whole of his assets without compensation. This is much more vicious than nationalisation, because the Constitution of the Commonwealth provides that if the Government desires to take over industry it can do only with proper compensation to the owners and shareholders in that industry. This coal legislation discloses the hidden hand of socialism, and surely in democratic Australia we cannot submit to confiscation without regard to the right of ownership."[370]   The final Coal Industry Acts did provide for fair compensation for any producer which suffered losses as a result of the new Board taking control of a mine. However the producers' demand for industrial relations to revert back to the Arbitration Court was not acceptable to the Commonwealth. The producers' demands in relation to mechanisation would in part be forced on them by the new Board, but on the key question of mechanisation of the process of removing coal pillars (the 'second workings'), this would take several more years to be resolved.

The Federal Opposition opposed many aspects of the legislation. Robert Menzies, then Opposition Leader, was extremely critical of the Bill in the Second Reading Debate in July 1946, referring often to the Davidson report and how he saw the Government's proposals in the Bill ignoring what Davidson had recommended. In relation to the new Coal Industry Tribunal Menzies made some observations that would prove over time to be pertinent. He said that the legislation had two great disadvantages: "The first is that …it completely severs

industrial arbitration in the coal industry away from the authority of
the Commonwealth Arbitration Court. It is unsound to whittle away
the authority of the …Court in this way and to pave the way for such
differential treatment of the coal industry as to encourage militant
coal miners to regard themselves as a race apart. It is unsound thus to
make a serious difference between coal mining and other industries and
particularly between coal mining and other forms of mining..." [371] The
second disadvantage according to Menzies was that "the authority of
the coal tribunal is seriously diminished by a provision in the bill for the
creation of local coal authorities and conciliation committees which, in
practice, will be presided over by lodge officers, who will be members of
the miners' federation, a practice which was very severely commented
upon in the report of the board of inquiry." One of the senior executives
in the industry who did not express much optimism about the new
arrangements was Thomas Armstrong, the JABAS chairman and colliery
proprietors' association chairman, who in his address to his company's
annual general meeting in November 1946 said that: "On past experience
one cannot see that the creation of a Board will give the much needed
increase in production, without the co-operation of all employees, but as
the Commonwealth Government proposes to spend large sums on the
Industry, some good may come eventually."[372]

With the NSW and Commonwealth Coal Industry Bills having
received Assent, the two governments soon announced the appointment
of the Board Members who would run the Joint Coal Board. The first
Chairman was Keith Cameron, an experienced metal mining engineer
and senior executive, who was manager of the North Broken Hill mine
prior to his appointment. He was not a coal industry man and would
continue as chairman until 1950. The other two Board Members were
A E Warburton, a senior official from the NSW Treasury, and R P Jack,
the Production Manager for the Coal Commissioner and a former senior
inspector from the NSW Mines Department. The Miners' Federation
was critical of these appointments and in particular the appointment
of Cameron, saying that he did not "fulfil the condition of being an
independent man…He's definitely linked with the big group of employers

in the mining industry. He had no experience in coalmining."[373]

However the two Governments were wise to have appointed directors who were not representative of the mining companies or the unions. That style of representation would come later in the Board's life, but in its early years, it seems appropriate to have kept the employers and unions at arm's length from its operations. The first three appointments gave the new Board a sensible blend of mining experience, senior management experience and financial experience, and importantly no ties with or obligations to the producers or unions.

The Joint Coal Board makes its presence felt

The Joint Coal Board commenced operating in March 1947. The NSW coal industry which was now to be managed and directed by the JCB was in a mess. The Board made it clear that the industry was unable to produce enough coal and was riddled with bitterness between the companies and the miners which was arguably unparalleled in any other industry. To achieve the necessary restructuring and stabilisation of the industry, there was an urgent need for more trained technical and managerial people.[374] While the JCB recognised that there were some mines that were efficient and equipped with modern machinery, it also knew that the industry overall was inefficient and out-of-date, with many mines needing drastic modernisation, and some mines needing to be closed down.[375]

The Board's charter and powers were extremely broad, but it decided, no doubt with the concurrence of the two governments, that it needed to focus on three fundamental objectives. In the Board's own words: "The Board considers that the fundamental objectives of the charter assigned to it by the Coal Industry Acts of 1946 may broadly be regarded as threefold: to provide sufficient coal …. to meet the requirements of Australian industry; to conserve coal resources and to ensure they are used to the best advantage; and to ensure that Australia is provided with its basic industrial fuel at the lowest possible cost … All of the other functions more specifically defined in the …Acts (such as… those relating to the welfare of mineworkers and their families and communities) should properly be regarded as subsidiary to these three basic objectives.'[376]

The JCB made it clear that it believed that, quite apart from the huge losses in production caused by industrial disputes (averaging 1.5 to 2 million tons a year), the industry was not in a position to supply the quantities of coal needed throughout the country. In fact it estimated that the shortfall was around 1 million tons, or around 12% of annual production. The Board also saw demand for coal growing strongly as the economy recovered from the war, with industries and households getting back to their normal lives and demand for new housing very strong. To eliminate the coal shortage it believed that NSW production would need to increase by around 50% in the period to 1953.

The JCB's immediate priority was to increase production as quickly as possible. It recognised that the prevailing shortage of coal meant that it could not afford to close down a number of inefficient collieries with high operating costs.[377] Rationalising the industry by closing such collieries would have to wait until later in the 1950s, when changes in the market would see collieries closing and employment numbers drop. While there were some mines which were equipped with modern machinery and which operated efficiently, the JCB saw the industry as fundamentally inefficient and out-of-date, with many mines having working conditions which it described as primitive.[378] It also recognised that the living conditions of many workers were primitive, and programs to improve these conditions would also be a feature of the Board's activities. But mechanisation was to be the key to driving the industry forward, together with major investment by the Board in developing its own production capacity in new underground and open cut mines.

The first year of the JCB's operations saw a flurry of activity, with an Order in July 1947 to NSW mines to install power boring machines by the end of October.[379] This was the first step in the Board's mechanisation program and was seen by the Board as a means of enabling miners to increase their daily rate of production relatively quickly. In 1947, of 127 underground mines only 23 had the full complement of power borers, and a number of other mines had a small number.[380] Boring machines were used extensively overseas and were designed to mechanise the old established practice whereby miners using picks would dig holes in the coal seam into which explosives would be placed.

While there was broad industry action to carry out the order, there were delays due to post war shortages and the lead times in securing imported equipment. The October date was not met by many mines, but the process proceeded and was complete in all the larger mines by 1948, and largely complete throughout the industry in 1949. A number of small mines employing less than 20 workers were exempted from the order. The mandatory installation of power borers did lead to some increase in production, but did not live up to expectations. The Board sheeted the blame for this onto the Miners' Federation, which it accused of "not generally (honouring) its undertakings."[381] The early benefits of the move to power borers were not evident in the first year to JABAS, with Thomas Armstrong telling the company's annual general meeting in November 1948 that: "Although the installation of Power Boring machines, as directed by the JCB, has been proceeded with, the output per man has not shown any appreciable increase, and so far had been very disappointing."[382]

In August 1948 the JCB advised coal companies of its plans to purchase sufficient machinery and equipment to fully mechanise NSW underground mines, not including pillars (the use of mechanical equipment to remove pillars was still subject to the legislative ban imposed by NSW Mines Minister Baddeley in 1941).[383] Companies were told to advise the Board by the end of September what their needs were to achieve full mechanisation. From discussions between the Board and the coal companies and a survey of the plans of all collieries, it became evident that many of the companies were unwilling to commit to the sort of expansion and modernisation program which the Board was proposing. In fact the survey found that, with a few "outstanding exceptions", no plans for future development of mines had been developed by the companies, and the majority also were not contemplating developing any such plans. There were several reasons for this lack of

interest or lack of action on the part of the companies – a lack of capital,
a lack of appreciation of the need for development and expansion, and
a belief that there was no point in investing in modernising mines if the
industry's industrial disputes record was to continue.[384]

For these reasons, and because of the expected delays in securing
new machinery and equipment (much of which would have to be
sourced from overseas), the Board therefore decided to develop its own
estimates of what was needed and to go ahead to place the necessary
orders. The Board's orders included coal cutting machines, coal loaders,
shuttle cars, conveyor equipment, electric motors and open cut mining
equipment and its first orders were in place by July 1947.[385] The initial
plan was that most collieries would purchase the necessary machinery
and equipment from the Board's pool, with only some being available
for hire where collieries lacked the necessary financial resources. By mid-
1950 over 50 collieries had obtained machinery from the pool, about
half of which was hired.[386] The Board clearly felt that it needed to push
ahead without delay and that doubts about the financial situation in the
industry and the ability of companies to pay were no longer obstacles in
the way of modernising the industry.[387] The intention of the Board was
to sell machinery at cost price to the companies, although it recognised
that some companies would need financial assistance.

Queensland goes its own way

In 1946, both the Prime Minister and the Miners' Federation pushed
the Queensland Government to become part of the joint NSW-
Commonwealth coal industry arrangements, but the Queensland
Premier, Edward Hanlon, wanted no part of such a body. He said the
he was "determined not to hand over the future development of the
industry in this State to a body outside the State. We dare not allow
competitive interests outside Queensland to control our industries,
because all industries are really dependent on coal. If you control coal you
control everything."[388] There had been a proposal to have the chairman
of the JCB, Keith Cameron, chair a Queensland coal board. But Chifley
then proposed a five person board, with Queensland having only two

members. Hanlon rejected this and said that Queensland would set up its own board and it would have "all the powers necessary to compel the mechanisation of the coal mines, and compel the adoption of modern conditions of labour."

Hanlon certainly won the support of the influential Courier Mail newspaper, which, like the Premier, was no doubt reading the minds of the Queensland voter. The newspaper called the Government's decision to go its own way a "victory on coal" and congratulated Hanlon "upon making a successful stand for Queensland's right to use its coal wealth for the State's advancement". It went on to say that "If Queensland can become a large exporter of coal, economically won by open-cut methods, it will be no longer possible for New South Wales coal miners and coal owners to retain their monopoly of coal supply to other States and to Australia's key industries. This monopoly has undoubtedly been abused in recent years."[389] The Courier Mail also gave the local members of the Miners' Federation some advice, warning them that "It can be expected that the Miners' Federation will be prompted by New South Wales interests to do all it can to hinder Queensland's entry into the inter State coal trade. Queensland miners who are members of the federation will be very short-sighted if they allow themselves to be used for this purpose. If they consider their own future, and the future of their children they will help Queensland to make the most of its great opportunity to expand production, enlarge its commerce, and acquire new industries as ensuring prosperity for a growing population." This interstate rivalry and the resentment by Queenslanders of control and influence from Canberra and the southern states would be a feature of other battles in the years ahead.

The Queensland Coal Board commenced operating in January 1949. Its powers, although more modest than the powers given to the JCB, were still strong and included: ensuring enough coal was produced to meet the State's requirements, ensuring that coal resources were conserved and developed appropriately, ensuring distribution and prices were appropriate, promoting the welfare of workers in the industry, and encouraging cooperation between management and workers.

Expert review of Queensland industry

In 1948 the Queensland Government commissioned a British firm of mining engineers and consultants, Powell Duffryn Technical Services (PDTS), to undertake a major comprehensive review of the Queensland coal mining industry. PDTS's report was finalised in July 1949. Davidson's 1946 report had presented a picture of an industry in very poor shape in Queensland, but given the controversial nature of the report, and in particular its complete rejection by the Miners' Federation, it did not carry the authority which it perhaps deserved. However the PDTS report, focusing on Queensland, being from an independent organisation, and compiled by specialists in geology, mining engineering and other fields, was much more influential, providing the framework and the detail for the QCB to develop its own plans to steer the industry into the future.

The PDTS report was in fact an indictment on the Qld industry for its insularity and backwardness, poor safety standards, inadequate capitalisation, poor management and poor regulation. In relation to the underground sector, PDTS noted that only 1 mine was fully mechanised, 1 other mine used mechanical loading for some of its production, only 3 used coal cutting machines, 12 mines used a total of 35 electric drills, only 7 mines used pneumatic picks, only 6 mines had washeries as part of their operations, and only 1 mine used a system of self-emptying skips.[390] There were 5 operating washeries, all in the Bundamba area, and 1 in Ipswich was under construction. All of these washeries were of a "rather primitive construction and design", and none of the washeries was found by PDTS to be efficient.[391]

Considering that mechanisation had started to be adopted in the industry from around 1905 in the Ipswich area with the introduction of coal cutting machines, the situation in the immediate post war era was shambolic. The industry's state was the result of neglect of the industry by governments over many years, including the Coal Boards established in 1933, and of course by managers and owners content to operate small mines which could reasonably be called "rat holes", with many of those owners and managers totally unfamiliar with the need for modern machinery, mine plans etc. In 1948 there were 91 privately owned mines in Queensland, producing an average of just under 15,000 tons per year each. The 4 open cut mines produced a total of around 235,000 tons, and the 3 State mines (Collinsville, Styx and Mt Mulligan) almost 210,000 tons in total. Collinsville was by far the largest mine in the State, producing over 170,000 tons.

PDTS noted that most of the seams being worked in the State were shallow and easily accessed by firms without the need for major capital expenditure. It was this latter factor, together with the ability of owners to develop small new mines with little commitment for machinery and equipment that had led to the turmoil in the industry in 1933 when the Coal Boards were established under State legislation. However it was clear that not much had changed in the intervening years. PDTS said that before the war, the increase in the numbers of collieries led to collieries working significantly fewer hours per week and to higher costs and poor profits. During the war, production was able to be increased to meet the demand, but quality deteriorated, and as companies had to focus all their efforts on production and with shortages of capital equipment, "the operators were still unable to put profits back into the industry."[392]

On a positive note, PDTS did find however that the average output per man shift (OMS) in Queensland's underground sector was not markedly different to level in NSW – around 3.5 tons vs 4 tons – despite the much higher rate of mechanisation in NSW. PDTS put this down to the employees of the smaller Queensland mines having a greater sense of "personal interest and responsibility".[393] PDTS also noted that Queensland had the distinction of being the only state up to that time which had not experienced serious shortages of coal.

In relation to safety, PDTS compared the fatal accident rate in Queensland with the rate in the UK coal mining industry for the period 1938 to 1947, finding that the rate in Queensland was 50% higher, and was also trending higher. The consultants recommended that a comprehensive mine safety organisation to encourage safe mining and to undertake research should be established and funded by the mine owners. While the adoption of modern mechanised methods of working would improve the industry's safety record, it cautioned that active supervision and research would need to be undertaken by this organisation.[394]

The standard of lighting in the industry was poor, and while PDTS noted that the Queensland Act did not require general lighting to be installed anywhere in a mine, it suggested that this was desirable and should be required in certain circumstances. For personal lighting, miners were almost all using carbide cap lamps which were fuelled by acetylene gas, with safety lamps only in use in certain conditions. The report did acknowledge however, that recent changes to regulations had now prohibited the use of these carbide lamps. Ventilation was also found to be generally poor, with inadequate air flow, and where fans were installed their efficiency was found to be low. PDTS recommended that the practice of turning off ventilation when mines were idle should generally be discontinued. Dust was another key issue examined, and one of the most damaging findings in their report was that there was "a very definite explosion hazard in many Queensland mines". In relation to the problem of pneumoconiosis among miners, PDTS concluded that the real incidence of this terrible disease was not known. It was believed that recently enacted regulations on dust suppression would lower dust problems, but PDTS warned that the regulations would need to be rigorously enforced.[395]

Among a range of other issues examined, PDTS also focused on the drilling which had been carried out over the years by producers and the Department of Mines. It concluded that drilling had been inadequate and not related to any overall plan of development. Detailed recommendations for drilling programs in key coalfields, including the Bowen Basin, were made, with these recommendations in part soon reflected in a burst of activity by the QCB and the Department. However the Bowen Basin's potential was not a priority for the Government in the next decade, and would only become more important for the Government after the ground-breaking discoveries by Thiess and Utah in late 1950s and early 1960s.

The Queensland Government and the QCB now had a detailed set of findings and recommendations on the basis of which the QCB could move the industry forward. However given the industry's size, limited financial resources and technical backwardness, this was a process which would take the best part of the next 20 years.

Blair Athol report stimulates interest in Queensland

The PDTS report was a wake-up call to the Queensland companies and Government and drove home the massive task of bringing the industry in the State up to a reasonable modern standard. Overall, it conveyed a very negative picture of the industry, although it also pointed the way to future development and the need for a huge program of drilling to define the coal resources known to exist. One potential bright spot in the late1940s however was the Blair Athol coalfield, around 385 kilometres by rail west of Rockhampton. The Blair Athol deposit was discovered in 1864 and the first mining began in the 1890s. Unlike the coal found in the Bowen Basin, Blair Athol coal is steaming coal, a far less valuable product tonne for tonne.

After World War Two, two companies were mining coal at Blair Athol, producing a modest 170,000 tons per year. The potential of the field had generally been understood to be significant for many years, but in 1947 the Queensland Government's Coordinator General, Mr J R Kemp, assisted by a technical team, undertook a detailed investigation of the field, including the level of funds needed to lift production and the rail and port implications of a greater level of production. Blair Athol was selected for examination based on the quality of the coal, the massive thickness of the seam, and its suitability for open cut mining. The only other field at the time which could have been considered for major open cut mining was Callide, but Kemp said that further investigation of Callide would be necessary before more detailed assessment was warranted.[396]

Kemp's report stated that Blair Athol was the largest known deposit of black coal in the Southern Hemisphere capable of being worked by open cut methods. With a seam ranging in thickness up to 28 metres, and a low ratio of overburden to coal, the potential for the field should have been good. One of the major problems was the lack of modern port facilities. Blair Athol was connected to the coast by rail, with links to Rockhampton, Port Alma and Gladstone, but each of these ports lacked modern loading facilities. The report concluded that the distance of Blair Athol from the coast and from coastal ports to interstate markets meant that the only viable opportunity for major development would be for the mine to sell to export markets. It recommended that the State Government approach the Commonwealth Government to seek its support for the development of the field, and that the Commonwealth and State Government representatives overseas be asked to promote the report to interested parties.

At around the time the report was being finalised, a proposal from British interests was made to the Queensland Government to develop production from Blair Athol of around 3 million tons per year, with a new railway line to the coast, and a new port, probably near Proserpine. The proponents planned to sell around 2 million tons to Asia and the balance to markets in Queensland and other States. The Queensland Government passed special legislation which would give the company rights to mine a large part of the coalfield. The Miners' Federation was bitterly opposed to the British proposal for development of Blair Athol, with a production level which was way above the current total annual state production of around 1.8 million tons a year, and employing 600 men compared with the roughly 3000 coal miners then employed state-wide. Federation president, Idris Williams, said that a project of this size would lead to cut-throat competition and conditions similar to the era between the late 1920s to late 1930s when miners could only get three days' work a week. He said that the project should have been taken over by the Federal and Queensland Governments, and not by 'British imperialists' who would make a huge profit and then go back home.[397]

The possibility of Commonwealth assistance for the development of Blair Athol became tangled in negotiations for possible Queensland involvement in the Joint Coal Board or a coal board which also involved the Commonwealth. With the Queensland Government deciding to set up its own coal board free of Commonwealth of NSW involvement, the Commonwealth's attitude was that the cost of developing coalfields and new mines in Queensland should be borne by the companies and the Queensland Government.[398] The British company carried out prospecting work on Blair Athol and work on a possible location for a route for a new rail line. It then assigned its interests in the project to another company.[399] However the proposed development did not proceed and this would be the first of a number of false starts for large scale development of Blair Athol, until the 1970s, when Rio Tinto (then CRA) finally made the decision to commit the investment necessary to develop the project.

Coalcliff Colliery 1964. NAA A1200 L48938

Hand mining Newnes Colliery Lithgow 1910. (Courtesy Mine Super)

7

Post War Industrial Relations

Miners win gains in wages and conditions

As we saw in 1938 and 1939, mine workers made significant wins in their campaign for better wages and conditions, first with the NSW Government's agreement in 1938 (supported by the Queensland Government) to appoint two Royal Commissions to investigate health and safety in coal mines and the case for a miners' pension scheme, and with the with the Drake-Brockman interim award decision in 1939 and the subsequent final award decision in 1940 by the full bench of the Arbitration Court. The new coal award gave underground workers a 40 hour week over 5 days, eliminated Saturday work, gave underground workers paid 10 days paid annual leave, and surface workers 11 days paid leave. Workers in other industries did not receive a 40 hour week until July 1947 (for those under NSW awards) or January 1948 (for those under Commonwealth awards). In line with the Royal Commission recommendations, the NSW and Queensland Governments also established new Miners' Pension Funds which commenced operating in 1941-42. The NSW Government also overhauled the Coal Mines Regulation Act following the Royal Commission on health and safety, although the new Bill was not passed until 1941.

Wages in Australia during the war improved initially, but then rose only marginally until after the war was won. From early 1942 wages were subject to wage controls, known at that time as wage pegging. On the day that Japanese aircraft began deadly bombing raids on Darwin, 19 February 1942, Prime Minister Curtin announced a broad range of controls over wages, prices and profits. Wages were fixed at levels prevailing on 10 February, and prices were now controlled by the Prices Commissioner so that the return on capital by business would be restricted to 4% (with profits in excess of this level to be subject to income tax), and absenteeism without a legitimate reason was banned.[400]

In NSW the average weekly wage for adult males in all industries (including manufacturing, mining, transport, building, agriculture and some other industries) rose from $9.66 in 1939 to $11.82 in 1942 and to $12.18 in 1945, an increase of around 26% over the six years. For the mining and quarrying industries in total, the average rose from $11.31 in 1939 to $12.55 in 1942 and to $13.73 in 1945, an increase of around 21%. For the coal mining industry, the increase in the average wage per employee over the six years was 21.5%.[401] While it is difficult to assess the detailed trends in coal industry wages during the war, with many miners working under the contract system, the wages for wheelers in the NSW northern district give an indication of the trend for non-contract workers. Wheelers' wages per day rose from an average of $1.90 in 1939 to $2.51 in 1942, after which wages rose only slowly to around $2.60 in 1945, giving an overall increase for the war years of about 37%. Up until 1939, there had been little movement in wheelers' wages in the previous 8 years, with the average daily wage in 1931 at around $1.80. From these figures we can assume that wage increases for non-contract workers were greater than for contract workers during the war years.

The years just after the war saw wages start to increase once again as controls were eased and increases began to be granted in the basic wage and in various awards by the Commonwealth Arbitration Court and other tribunals. The Central Coal Authority, A C Willis, granted contract miners wage increases in March 1946, but this decision was overturned by the Coal Commissioner in May. In October Willis awarded coal workers an additional week's paid annual leave, and also granted 5 days sick leave per year. The secretary of the Southern Colliery Proprietors' Association, William McNally, said that the decision gave the members of the Federation far better conditions than miners in other countries, and put them above other workers in Australia, with underground workers now earning a minimum of $14 per week plus the 3 weeks' annual leave and one week's sick leave.[402] Given that average weekly adult male wages in NSW in 1946 were around $13, and most workers only enjoyed only two weeks' paid annual leave, coal mine workers had moved ahead of workers in other industries, although not markedly.

In 1946 unions in a number of industries began a serious push to gain significant wage increases, and by October a campaign backed by the ACTU was demanding an increase in wages of $2 per week. These demands were not successful, but in December 1946 the Arbitration Court granted a basic wage increase of $0.70 per week. This increase flowed directly to industries under Federal awards, and as a result, wages in mining and quarrying in NSW rose by around 7%. This was the first substantial wage increase for many workers since early 1943. In December 1946 there was also a major change in regulations, with industrial authorities and tribunals now permitted to increase wages in some circumstances without the need for approval from the Arbitration Court, or in the case of the NSW coal industry, the Joint Coal Board. Shortly after, in April 1947, regulations were again changed, a move which freed up the system even further, giving industrial tribunals the power to increase wages provided the Arbitration Court (or the Joint Coal Board) deemed that any change was not detrimental to the national interest.

For the workers in the coal industry, it was 1947 that saw the start of significant further gains in wages and conditions. In April 1947 the newly established Coal Industry Tribunal (CIT) granted workers payment for 11 statutory holidays per year; in May 1947, the CIT followed up with an increase in annual leave from 10 to 15 days, and an increase in sick leave from 5 to 10 days per year. In November 1947 the CIT was able to give the contract workers in NSW, Queensland and Tasmania the $0.30 per shift increase that had been rejected in May 1946, and with the increase applying from August, the miners received a welcome Christmas bonus. CIT chairman Frank Gallagher also made it clear that he believed that the coal industry was able to absorb wage increases, and that it was important to maintain the supply of coal to key users. At that time, the NSW coal industry was still being subsidised by the Joint Coal Board, and wage increases tended to be reflected in the level of subsidy paid.

In April 1948, Gallagher replaced the provision in the awards for employment in the industry to be on a daily basis with employment on weekly basis. This decision, based on an agreement with the producers and the union, was a major win for the Federation, with the its general president Idris Williams acknowledging that it represented "an important gain in improving miners' working conditions."[403] The Queensland Times reported that the decision had been described by union officials as the greatest gain in working conditions yet made by the union.[404] The decision meant that the producers needed to provide work five days per week for those employees who attended for work and who were willing to work. When work was unavailable, the producers would now be required to pay workers a full days' pay at the minimum district rate. Management retained the right to deduct wages for employees who had been stood down for example for misconduct or refusing to carry out their duties; and in the event of a breakdown in machinery, mine workers were now to continue to receive wages for four days. Williams noted that while workers needed to understand that if they were unable to carry out their normal duties due to a breakdown, they were obliged to do the work reasonably requested by management. However he also said that the new award meant that employers could no longer operate "as they desired", standing down employees or even sending them home without paying them. Up until that decision, with employment on a daily basis, workers who presented for work and who were not required that day received two hours' pay. In March 1949 Gallagher made more major changes to the coal award, increasing pay for overtime for members of the Federation in NSW and Victoria through a 50% loading for the first four hours, and double time thereafter. The loading for afternoon shifts was increased from 7.5% to 10% and for the night shift from 7.5% to 25%. These changes followed changes to the awards applying to miners under the metal trades awards.

Workers in the coal industry had won some significant gains in wages and conditions since the end of the war. However if we look at the relative levels of average wages in NSW in mining and quarrying (a sector which was dominated by the coal mining industry)[405] compared with the all industry average, between 1939 and 1949 the mining and quarrying sector's average increased by around 55%, and the all industry average by around 77%. In 1939 the mining and quarrying average weekly wage was around $11.30, 17% above the all industry average. By 1949 the mining and quarrying average was around $17.50, only around 2% above the general industry level. Coal industry average wages and salaries, approximately $11.10 per week in 1939, rose to approximately $18.40 per week in 1949, 7.5% above the all industry average.

The coal mining industry of course had won greater paid leave entitlements, and in 1949 also won a long service leave scheme (the commencement of which was delayed for most workers until 1950). These benefits meant that the effective margin between coal mining and other industry wages was greater than the weekly average figures suggest. Employees throughout Australia had won significant wage increases after the end of the War. The coal industry had shared in these increases. The next few years would see coal industry wages surge, but not before the most damaging strike to occur in the post war period.

A strike like no other

The national coal strike of 1949 was one of Australia's most significant industrial disputes; it was significant for the action taken by the Chifley Labor Government to break the strike, for its impact on the fortunes of the Labor Party for many years, and for its impact on the economy at the time. The disputes which had plagued the industry during the war, and which continued to occur after the war, reduced the production of a much needed commodity and caused significant problems for other industries and for long-suffering households. Throughout a good part of the 1800s and the years leading up to World War Two, the coal industry had the distinction of being by far the major source of industrial disputes in Australia. In 1929 and 1930, reflecting the northern lockout, approximately 4.3 million working days were lost in the NSW mining industry, the great proportion of these in coal mining. All other industries combined in NSW recorded around 0.8 million working days lost for that period. For the years from 1931 to 1939, mining recorded around 2.2 million working days lost, almost 70% of the total number lost by all industries in NSW.

By mid-1948, with production unable to meet the demand for coal, the situation in NSW had deteriorated with coal stocks severely depleted. On 22 June the JCB announced drastic cuts to coal allocations to power stations, gas plants, railways, steel mills and other factories. Coal supplies to power stations and gas plants were cut by 25% and to the Newcastle and Port Kembla steelworks by 15%; general industrial users saw an average cut of 30% and the railways a cut of 5%. Allocations to other States were also cut back. The following day the Sydney Morning Herald's front page announced the restrictions and also the news that use of electricity was banned in most factories that day, with employers estimating that 100,000 workers would lose a day's pay. The ban was imposed by the NSW Government under an emergency order after the city's electricity supplier (the Sydney County Council) advised that its coal supplies were almost exhausted. The Council said that it was using "the scrapings of its coal bins to fire the boilers at Bunnerong", the major power station in Sydney at the time.[406] The central executive of the Miners' Federation issued a detailed statement denying that its members were the cause of the problem, pointing to a modest 50,000 tons increase in coal production for the first half of the year compared with the same period in 1947. This had occurred said the Federation at the same time as newspaper "headlines were screaming out their vilifications of the miners" and that the plain fact was that the coal industry was "not equipped to meet the increasing demands being made upon it". The Federation also claimed that the miners were being made scapegoats by the coal producers and that the shortages were due to the incompetence of the producers and the failure of the JCB to reorganise the industry.[407] One thing all stakeholders in the industry could agree on was that the coal industry was unable to meet the growing demand for this vital commodity.

The JCB was concerned about the shortages of coal, but also saw a need to re-build stocks to provide a buffer against future disruptions. The NSW Government on the other hand was concerned about what it saw as an inappropriate JCB plan to build up stocks in the middle of winter and it forced the JCB to relax the restrictions.[408] After several days' negotiations involving the JCB, the NSW Government and the unions, the JCB announced a relaxation of the cutbacks to power stations and gas plants, reducing the cutbacks to only 10%. However the cutback in the steelworks' coal supplies was now increased to 20%. By November 1948 the Board was forced to act again, ordering the most severe rationing ever seen in NSW. The steelworks' supplies were reduced, this time to only 10% of normal, and supplies to railways were slashed by 66%, electricity power stations by 40% and gas works by 50%. These cuts caused massive disruption to industry, including major retrenchments in the steel industry; prospects for Christmas 1948 were looking particularly bleak. The strikes and the rationing periods in 1948 caused losses of wages, cutbacks in production by many key industries, and severe, even if short lived, impacts on families. With the Federation leaders being blamed for the turmoil in the industry Governments and the public would now be even angrier when the national strike in 1949 commenced. [409]

In 1949, the Miners' Federation lodged a log of claims with the Coal Industry Tribunal involving an application for long service leave, a 35 hour week and a 30 shilling ($3) per week wage increase. The 35 hour a week claim was adjourned in May at the request of the union. The claim for the wage increase was withdrawn shortly after it was lodged. The Tribunal completed its hearings on the long service leave claim on June 9 and was planning to grant the claim on June 14. However the union met between those dates and decided to commence a general strike on June 27 unless all its claims were agreed to. The Tribunal decided to hold back its decision because of the union threats. The union then went ahead with the general strike. The 1949 strike lasted for around seven weeks, commencing on June 27 and ending on August 15.

Judging by newspaper reports, the public was strongly opposed to the strike, and this is understandable in view of the strike's impact on jobs and everyday living, and also in view of the communist threat which was a major post war issue. While the strike was relatively brief in comparison with some other coal industry strikes or lockouts (eg the fifteen month lockout on the northern NSW coalfields in 1929-30), it had a serious impact on an economy recovering from the war. Hundreds of thousands of workers lost their jobs or were laid off as factories could not obtain sufficient electricity or gas. Rail transport was severely affected and unemployment levels rose to the highest levels since the Depression. It was winter and there was little or no gas for heating, households had to rely on candles for light in their houses and some people were forced to cook in their back yards.

The communist threat in Australia had been building as a political issue since the end of the war. It had also been an issue early in the war, with the Menzies government declaring the Communist party of Australia an illegal organisation in June 1940. However following the German invasion of the Soviet Union in June 1941, the Soviet Union was now on the side of the Allies, and communism became much less of an issue until after the defeat of Germany in 1945. Communist leaders were in senior positions in key industries in Australia after the war, notably in coal mining and on the waterfront. This, combined with the Soviet Union's territorial expansions at the end of the war, the influence of communists in countries such as France and Italy, and the increasing ascendancy of Mao's communists in China, meant that many people in Australian saw communism as a real threat to their way of life in the post war years.

Winston Churchill, the British wartime Prime Minister, in his famous 'iron curtain' speech in the USA in March 1946, had sent shivers down the spines of many Americans, and no doubt Australians, when he said: "From Stettin in the Baltic to Trieste in the Adriatic an iron curtain has descended across the Continent. Behind that line lie all the capitals of the ancient states of Central and Eastern Europe. Warsaw, Berlin, Prague, Vienna, Budapest, Belgrade, Bucharest and Sofia, all these famous cities and the populations around them lie in what I must call the Soviet sphere, and all are subject in one form or another, not only to Soviet influence but to a very high and, in some cases, increasing measure of control from Moscow."[410] Writing about the decision to strike, Fred Daly, a long serving Federal MP and Minister in the Whitlam Government, said that "It was apparent that the Communist Party was determined to make its play at this time on these issues and it was precisely what it did. The Miners' Federation called a general strike in support of their demands. "[411]

When the miners had been threatening to strike, Chifley had conceded that conditions in the mines had been intolerable, but also referred to improvements which he said his government had made since the end of the war.[412] However once the strike began Chifley took a hard line and refused to negotiate with the union. Chifley's biographer David Day wrote that the decision to strike "challenged Chifley's abiding attachment to the system of arbitration, especially as he had established this tribunal himself. It also challenged his attempts to keep inflation under control and to some extent represented a political challenge to the Labor Party, with communists being strongly represented among the miners' union officials and the Communist party coming out strongly in support of the strike. "[413]

The Federal Government ran large advertisements in the press on 7 July saying: "This is a strike against arbitration. The Commonwealth Government has told the Australian people, the rank and file of the Miners' Federation and trade unions generally what the facts are of the coal strike, so that any distortions of the truth by the Communist sections of the Miners' Federation will not confuse anybody's mind and the Commonwealth Government again makes the issues clear. This strike was planned months ago by the Communist sections of the miners' leaders. They do not want the claims to be dealt with by arbitration. Miners, do not be misled by the communists who want to wreck the arbitration system. "[414]

Chifley was positioning his Government in direct opposition to the Federation's leaders, and was determined to teach the communist leaders that they had gone too far. Edgar Ross, a senior member of the Federation and the editor of its Common Cause magazine, later wrote that he tried to negotiate a deal with the Government through the secretary of the Joint Coal Board under which there would be an in-principle agreement to a 35 hour week, with the shorter week only to be introduced when production was sufficient to meet the demand; in return the unions would accept rationalisation of the industry as sought by the JCB and the producers.[415] But, according to Ross, the Prime Minster was not prepared to do a deal, and it later became known that Chifley made his attitude clear to the Labor Caucus, saying that: "The Reds must be taught a lesson." Whether Ross was authorised by the Federation leadership to offer such a deal is not clear, but his efforts came to nothing, and the outcome was still an extended strike which caused massive disruption to the country.

The Government legislated to freeze union funds and made it illegal for anyone to donate to the union strike fund. When union leaders refused to divulge the location of union funds they had withdrawn from the unions' bank accounts, eight senior officials were sent to gaol for 12 months for contempt of court. They included five leaders of the Miners' Federation (Idris Williams, the general president; GWS Grant, the general secretary; W Parkinson, the vice-president; J H King, the secretary of the Western District; and M Fitzgibbon, the secretary of the Southern District). J Healey, the general secretary of the Waterside Workers Federation, and L McPhillips, the assistant secretary of the Ironworkers' Union, were also among those gaoled. The Miners' Federation was fined the equivalent of $4000, and the Waterside Workers Federation and the Ironworkers Union both $2000; several individuals were also fined.

The most dramatic move came on August 1 when, five weeks into the strike, the Government sent Army troops in to work some NSW open cut mines. According to Day, the Government also had plans for the Army to work in the underground mines, but "With its leaders in gaol and much of the trade union movement pressuring the miners to give up, the end was already in sight before the troops were brought in…" Day also noted that "Around Lithgow, walls and railway bridges were emblazoned with 'Ben Hitler' as the striking miners hit out at their local member."[416] Whether it would have been practical and safe to have the Army working underground is debatable given the safety issues and the time it would have taken to train soldiers in the many tasks which underground mining involves. Nevertheless, the fact that the Government was planning or even considering this option demonstrates its determination to break the strike.

The strike also had a huge impact in other states, particularly in Queensland, where most mines were idle. However one area in Queensland where a number of mines continued to operate when their employees refused to strike was the Rosewood area west of Ipswich. The small family owned and operated mines in Rosewood soldiered on, without the need for any military assistance. These mines included Normanton, Mt Elliott No.1 and No.2, Smithfield, Lanefield Extended, Westvale No.5 and Rosewood, and they were able to supply sorely needed coal to the Brisbane market. On Monday 12 July the Courier Mail reported that in the biggest movement of coal since the start of the strike, a convoy of 137 trucks, under police guard, carried almost 1,000 tons of coal from Rosewood to key gas and electricity plants in Brisbane the previous day. The chairman of the Queensland Coal Owners' Association, Mr S Trewick, claimed that 16 of the 18 mines in the Rosewood area had worked over the weekend, with the other two expected to resume work that day or during that week.[417]

However with the strike in full swing in NSW, the cracks in the union's front started to appear. Some of the mines in the Rosewood area which had been on strike did resume work in the first half of July and by the end of the month 38 mines in Queensland were back at work. In Western Australia, aggregate meetings of mine workers on 17 July decided on a return to work.[418] The writing was now on the wall for the NSW mine workers and the Federation's leaders. On 27 July the Combined Mining Unions Council, representing the various unions including the Federation, voted to hold aggregate meetings in NSW. On 28 July the Acting Central Executive of the Federation (acting because a number of the leaders were in gaol) rejected that decision, causing a walk-out from the Council by union leaders who did not support the policy of the communist Federation leaders. While the Federation's leaders held to a policy of refusing to hold mass meetings, meetings in the coal districts did occur and increasingly showed opposition to the strike. Also, some leaders of the Federation's northern district stated that they had not supported the strike from the start.[419] The strike in NSW formally ended when the Federation was forced to order a return to work on 15 August in what became a resounding win for the rank and file members of the unions.

The use of Army troops to mine coal in a number of Hunter Valley open cut mines is one of the enduring elements of the 1949 strike. However, as the JCB made clear in its second annual report, the troops' efforts actually contributed only modestly in terms of the amount of coal produced and distributed to users. The JCB worked with the Commonwealth Government to have the Army work on a 3 shift, 6 days per week basis, commencing on midnight 1 August to midnight 14 August. The coal produced totalled just over 100,000 tons.[420]

The 1949 strike had the potential to be a violent one, but perhaps the memories of the Rothbury Riot in December 1929 may have helped to avoid violence in 1949. In 1929 one miner was killed, and many others wounded or otherwise injured, when police clashed with striking miners who were demonstrating outside the Rothbury colliery near Cessnock. Whether the shot that killed the miner was intentional or accidental has long been debated, however this clash was one of the most violent in Australian history since the Eureka stockade battle, and would not have been forgotten by those involved in 1949. Some of the public meetings held during the 1949 strike however were extremely heated, but generally not violent. However one anecdote from Edgar Ross in his History of the Miners' Federation shows that violence can never be discounted. Ross wrote that "Indelibly impressed on my memory is the day when I had almost to leg-rope an enthusiastic supporter, an ex-member of the Irish Republican Army, to prevent him from dynamiting the Hawkesbury railway tunnel to stop scab coal reaching Sydney."[421]

The strike effectively ended the union careers of a number of the Federation's leaders and the communist influence in the industry was reduced for several years. However, while the leaders of the Federation must carry much of the blame for the strike, the early support from the rank and file in NSW should not be forgotten. In mass meetings in the coal districts at which mine workers voted on the Central Council's recommendation to hold the strike, the vote was around 8000 in favour to 800 against. At the northern district meetings in Kurri, Abermain and

Cessnock, the recommendation to strike was supported by 98.5% of workers. And following the decision of the Australian Railways Union to agree that its members should operate trains to move coal during the strike, mine workers voted to continue the strike, with the vote in Kurri 1000 to 3.[422]

Queensland dispute level better than NSW

In the year ending December 1946, the first full year after the end of the war, the NSW mining industry recorded 300,000 days lost to industrial disputes, 23% of the total for all industries. In 1945, mining accounted for 34% of total days lost. This was not a record to be proud of, and certainly not for the coal industry which accounted for the majority of those days lost in the NSW mining industry.[423] The year ending June 1950, which included all but 4 days of the national strike, saw the NSW coal industry losing 14.2% of possible shifts to industrial disputes, an average of 36 days lost for every employee in the industry. The year ending June 1948 saw 9.8% of shifts lost to industrial disputes, and an average of 25 days lost per employee.[424]

The record in Queensland for industrial disputes has traditionally been significantly lower than in NSW. In the year ended December 1949, 8.2% of possible shifts were lost to industrial disputes; given this was the year of the national strike, this was a surprisingly modest number. In the following calendar year, the rate had fallen to 1.7%. While the low rate in 1950 would be followed by a jump in 1951 (to 4.6% of shifts lost), the rate over the subsequent few years averaged around 2.5%, and fell to only 1.5% in 1955.[425] In contrast, the rate in NSW, while it improved significantly during the 1950s, did not dip below 4% until the late 1950s; the 1960s then saw further improvement, but this was followed by higher disputation in the 1970s and early 1980s.

The psychology of the striking miners

It is still puzzling today why the mine workers in NSW, particularly the northern district miners, were so strike-prone in the years during and after World War Two. Of course the coal industries in Australia, and in other countries (including the UK from which the industry has derived much of its heritage and culture), have also had a major history of disputes. But having won significant gains just before World War Two, and more gains during and after the War, why the terrible record of disputation during the War and in the first few years after it? While various theories can be put forward, it is interesting to look at how several authors have described the actions and motivations of the mine workers.

Colin Davidson, in his 1946 Inquiry report, treated the mine owners fairly lightly, but came down hard on the Miners' Federation for its "weak and divided" leadership and its attempts to impose discipline on its members which were "substantially a failure, although in a few instances they proved to be effective". He concluded that "Especially in the northern and southern Districts of New South Wales, discipline is almost non-existent amongst mineworkers who are members of the Miners' Federation and are within its sphere of influence". However he also noted that "Discipline is observed by mine workers who are not members of the Miners' Federation or are remote form its influence and generally, also, by mineworkers in mines that are free from the system of payment on contract rates."

Davidson then attributed the lack of discipline to a number of factors, including "weak and divided leadership" in the Miners' Federation, political antagonism between Communists and non-Communist members of the union, the fact that nearly all strikes achieved some success in gaining concessions, the "inability or reluctance on the part of the Government to enforce the law", appeasement by the Government in yielding to union demands, and the opposition by the Federation to the dismissal of employees by management "for any reason whatsoever". In relation to strikes during the War, Davidson stated that almost none was due to management impropriety, and that prosecutions under the wartime regulations should have been enforced more strongly as those prosecutions that had been made for absenteeism and other offences achieved considerable success.

The Miners' Federation, having withdrawn from the Inquiry and having dissociated itself from it, would have none of Davidson's conclusions. However it is interesting that a couple of years after the report, former Federation general president Harold Wells wrote an article for the Sydney Morning Herald at a time of major disputation and when the Joint Coal Board was in its infancy. The Herald's headline read: "The Miners Should Realise Their Responsibilities" and in his article Wells acknowledged the gains which had been made by mine workers in pay and conditions, but he stressed that so far in 1947, 1.3 million tons of sorely-needed coal had been lost to disputes. He asked whether this recent record was "evidence of a new attitude on the part of the mineworkers" and if it was, "then the mineworkers are going to miss the bus."[426] Wells went on to say that in his opinion "the miners of Australia are at the crossroads. Started along the new road is the organisation of the Joint Coal Board, with the money, authority, organisation, machinery, mining knowledge, and the will to rebuild the coal industry in line with modern times and ideas. Mr. Gallagher (chairman of the Coal Industry Tribunal) too, is on that road. The miners are partly along the other road- the hard, rutted roadway of blank days, depressions, isolation, bitterness, and criticism. They must make the choice soon, or it will become more and more difficult to make. Which road are they going to take?...To-day, as never before, the miners need labour discipline to ensure that the greatest opportunity that ever came their way to build a new kind of life is not frittered away… A strong and disciplined union in the next five years can be of great importance to the rebuilding of the Australian coal industry, and in this regard a service to the Labour movement and to Australia… If the attitude of the Australian mine workers is to remain one of irresponsibility and the chasing of pseudo-principles and outmoded and hurtful customs and practices, the union will go by the board, playing no useful part, and of only very minor importance to the Joint Coal Board."

Albert Willis, as chairman of the Central Coal authority, came out strongly in 1944 against the trouble being caused by what he said was a small section of people in the industry. Willis said that anyone who was doing the work of "fifth columnists" within the industry should be removed. Willis was dealing at that time with an application by the Federation for a new award, and announced that until steps had been taken to produce sufficient coal to relieve the serious shortage, he did not intend to proceed to deal with the application. While he said he was looking to the Federal Government to take whatever action was necessary to control the industry, he put much of the blame for disruptions on "elements within the Miners' Federation" who were continually refusing to obey their organisation. Willis urged those workers to seriously consider whether it was in their own interests to continue their present conduct. Perceptively, he said he was convinced that "the whole industry would need to be thoroughly overhauled and reconstructed before a permanent peace (could be) achieved".[427]

A very different perspective on the coal industry was provided by Alan Walker, who later became a minister and leader of the Uniting Church, and who spent two years after the war living in the town of Cessnock in the lower Hunter Valley. Walker wrote a Master's thesis on the mine workers and their families who lived in Cessnock, then a major coal mining town of around 15,000 people located on the South Maitland coalfield.[428] Walker surveyed people in the town, and, having lived there, developed a fair understanding of the way many thought. Walker saw four major landmarks in the development of the town: the change from a farming to an industrial community, the 1923 Bellbird Disaster, the 1929-1930 lockout, and the Depression. He said that "Each of these incidents not only belongs to the past, but lives today in the minds of the people." Walker found intense feelings of bitterness in the town: "Across the community and the minds of the people is a dark shadow; it is that which comes from the economic collapse of 1929 and the incidents which followed in subsequent years. …..All this means … that almost a pathological condition has been created. Mental attitudes have become so warped by suffering and disappointment that suspicion and bitterness are uppermost. It has estranged many from the Australian society as a whole and engendered a permanent spirit of antagonism toward their fellows."[429] Cessnock was a town with many poor families and very poor housing. The long lasting effects of the Depression are also reflected in a quote from one miner in Walker's report: "We were out of work for nine years, and had to come and live in this tin hut of two rooms. We have never been able to get out of debt until this year, and have had to remain on in this shack".

Other important aspects of Walker's work related to the effects of change on the workers in the industry, and in particular mechanisation, which was described by one national union leader as a "Frankenstein", with workers fearing for their jobs, and older workers, who used to be proud of being good miners, becoming disoriented by change. With many mines still to be mechanised, the union leader said that workers feared that many would be thrown out of work and communities disrupted.[430] The nature of the working environment underground was also critical in Walker's view. "To work underground means darkness, dampness, constant peril and bad ventilation. For many it involves much walking, sometimes three to four miles per day each way, from the bottom of the mine-shaft to the coal-face. This itself imposes its physical strain without actual labour. ….But by far the most important factor in this connection is the high accident rate in the mining industry."[431]

But why the incredible urge to strike in the coal industry, particularly in NSW? A study which gives some important insight is one by Andrew Metcalfe - *For Freedom and Dignity: Historical Agency and Class Structures in the Coalfields of NSW*. Metcalfe quotes Harry Cockerill, a former miner at the Richmond Main colliery, a large colliery near Kurri Kurri: "The last year I was at Richmond Main, 1948, there were 54 petty one-day stoppages in the year. Every one of those could have been avoided. The official Federation position was that those who caused the strike were wrong … and you tried to prevail on them not to do it….They used to have a saying there: a bloke would come home from Richie (Richmond Main) and the housewives would say 'what's wrong this morning?' He would answer, 'I don't know Missus, but we're not going back till we get it'….It was characters who caused the stoppages. They weren't Comms (Communists). They were just blokes that didn't want to work. …Irresponsible strikes were one of the mysteries: it happened because of the unpredictable loyalty that was amongst them. At Richmond Main nearly all the stops were caused by the wheelers, spare miners and road-layers. Most were young men, single men. It wasn't for political reasons."[432] Metcalfe also quotes Athol Lightfoot, a manager of Richmond Main: "A fellow at Richmond Main had the most remarkable ability to take the pit home …He could get up there and harangue these fellows, and they'd empty their water bottles out and they'd say to one another 'What are we going home for?' and they wouldn't know. They don't know whether they are voting for or against a resolution most of the time…(they) know it's going to cost them a certain amount of money, but most of them are lighthearted…The Lodge officials were defeated many many times…the Lodge officer in general had very little authority."[433]

While industrial disputation in the industry caused great damage to employees, their families and communities as well as of course to the producers, some disputes and strikes did have a humorous side to them. In 1942 JABAS was fined the equivalent of $100, and Athol Lightfoot, as the colliery manager, was fined $200, under the wartime coal security regulations for allegedly closing the Richmond Main colliery for two days in September. At that time approval from the Coal Commissioner would have been needed for the colliery to close, even temporarily. Christopher Jay, in his book *The Coal Masters*, drew on the family memoirs of Lightfoot: "A contract wheeler, who moved coal skips with a horse, objected to the chief ostler… transferring his horse to another job. The ostler's authority to do this was specified in the Blue Book (a book drawn up by one of the mine union leaders setting out conditions of employment and seniority). Nonetheless, there was a sit down strike, with 600 men just sitting on the shaft bottom each morning. After phoning the Miners' Lodge president …to warn he would blow the 'no work' whistle, Athol Lightfoot did just this on two occasions, earning him five continuous days in the witness box."[434] In the court case relating to this incident, the chief ostler was described as an excellent veterinary surgeon who had a great knowledge of horses and was in charge of 600 horses at a number of the JABAS collieries. In the case it was established that there were lists of horses which were allocated to specified periods, but it was also found that the chief ostler was unable to read. Jay also wrote that when the management of Richmond Main and the union officials were down in Sydney for the hearing, and another person was appointed to manage the colliery, there was not a single strike. "It became folklore that the path to industrial peace was to keep the management and the union officials away from the colliery."

But what about the role of the mine owners and managers generally? Surely mine owners and senior managers carry some of the blame for the industry's poor industrial relations record? As we have seen, Prime Minister Curtin clearly believed that the major cause of disputation during the war lay with the mine workers. However Justice Drake-Brockman who conducted the historic 1938-1939 coal award case was extremely critical of mine management and laid much of the blame for the state of the industry at their feet: "There appears to exist a constant tension combined with distrust between the parties. Employees have usurped some of the functions of management particularly in relation to employment. Managers and superintendents generally seem to lack the necessary courage to assert their rights and consequently the general state of discipline is bad. The 'last on first off' rule is rigidly enforced by the Federation and apparently regardless of such matters as misbehaviour or inefficiency. A miner with several years' seniority is so firmly entrenched in his job that nothing short of the closing down of the mine, death or a compensable illness can prise him out of it. He can stay away from time to time as he feels inclined, but when he returns to the mine his job must be waiting for him. 'Silly' strikes and sectional stoppages for insufficient reason and often for no apparent reason at all, are all too common. These practices are quite unjustified, they are harmful to the industry and to all people associated with it. The success and prosperity of a particular colliery are the concern of all employed in and about it. The employees' earnings and security of employment are directly involved. These things ought to be obvious, but unfortunately they are often disregarded."[435]

In 1948 JCB chairman Keith Cameron issued a long statement to the media in response to criticism of the JCB and its failure to overcome the chronic shortage of coal. Cameron said that he was "not completely inexperienced in industrial matters and, in fact, could claim wider experience than most" and went on to acknowledge that he knew that there were "extreme militants in the miners' union, just as there (were) extreme reactionaries within the ranks of the coal owners." Cameron recognised that the majority of the union members were "good citizens", just as there were "many liberal-minded men among the owners and their representatives, trying to help lift this industry out of its difficulties." He reasoned that "The difference in the public view is that the undesirable activities of the few in the union receive prominence, while those of their counterparts among the owners are seldom sheeted home."[436]

In December 1950 the CIT made another significant change to the coal award by introducing an attendance bonus of an additional day's pay for each worker where both the worker and the colliery had worked a continuous ten day fortnight. This decision enraged the Miners' Federation which arranged for pit top meetings at the various collieries to protest. As a result, after the Christmas holidays, when the mine workers had worked one day after their annual leave in order to qualify for payment of the public holidays, the meetings agreed to strike on a weekly basis. The CIT remained resolute and in March 1951 mass meetings voted to end the strikes. The impact of the bonus was interesting. In 1952 JABAS's chairman, Thomas Armstrong, reported to the company's annual meeting that over the last year wild cat strikes had been reduced, particularly towards the end of each fortnight, with workers reluctant to lose the bonus because of a one day strike. But JABAS also found that when strikes did occur they tended to be longer than before.

In chapter 5 the critical role of the contract system was noted, with Davidson highlighting it as one of the industry's major problems and constraints. The Joint Coal Board also zeroed in on the contract system in its early annual reports, as well as on the reasons for the overall poor industrial relations in the industry. The Joint Coal Board's first annual report for the 1947-48 year added another perspective on the industrial situation it had inherited, although it throws little light on the reasons for the actions of the striking miners: "It is probably true that the present industrial unrest in the industry derives largely from the bitterness and fears of the past and, to some extent, may be accentuated by unwise management, but, despite this, miners' leaders of all shades of political opinion have time and again condemned petty stoppages. It is clear that, at the very least, one million tons of coal is being lost a year through avoidable stoppages which… can have no possible justification…" In that inaugural annual report, the JCB also identified the inability of the Federation's leadership to control its rank and file as another major cause of disputation: "Experience has …forced the Board reluctantly to the conclusion that the miners' leaders are either unwilling, or perhaps unable, to control their own members and that not only are the undertakings of no value whatever but, considered objectively, they can only be regarded as tactical manoeuvres designed to meet a particular situation."[437]

In its annual report for 1949-50, the JCB confirmed what it had said in its first annual report in relation to the problems of the contract system, namely that employers and employees had long believed that the contract system, including its dargs (production quotas), cavils (ballots for work locations) and consideration payments (payments to account for difficult working locations), was a major factor in causing industrial unrest. The system also permitted some miners to earn in 3 or 4 days' work what other workers could earn in a full week, thus not constraining those miners from striking. The JCB also said that it was "unquestionably true that the incidence of strikes in mines operating …under the contract system is greater than those operating with day labour, and, in particular, the fully mechanised mines."[438]

The causes of the industry's woeful industrial disputation record during and after the Second World War no doubt lie in a complex web of factors. The troubled history of the industry, the culture inherited from the British industry by both managers and employees, the industry's inherent dangers and poor safety record, the 1929 lockout in the northern district, the Great Depression and the severe impact on employment, the recurring intermittency, the contract system of payment, the seniority system and the difficulty in sacking certain workers, management and union and worker intransigence, and perhaps downright stupidity at times all combined to produce a totally unacceptable situation. It would take a totally new organisation with sweeping powers to bring about the changes in an industry which was vital to the post war development and prosperity of the country. New technology and the mechanisation of the industry would also play major roles in the transformation of the industry.

8

Mechanisation the catchcry: the 1950s

Mechanisation in NSW now full steam ahead

The Powell Duffryn report on the Queensland coal industry, outlined in chapter 6, put that state's shortcomings and challenges very clearly. For the NSW industry, the JCB's first annual report made it crystal clear that it was also in a poor state in terms of mechanisation: "The industry is fundamentally inefficient and out-of-date. While there are some properly equipped and efficiently operated mines, many need drastic technical modernisation whilst others should be closed down."[439] While many NSW companies were well ahead of Queensland in terms of mechanisation, and some had state of the art machinery and equipment, there was an urgent need for all producers to modernise and become more efficient. Two of the key indicators of mechanisation adopted by the Department of Mines and the JCB were the proportion of coal mechanically loaded and the proportion of coal produced by coal cutting machines. Mechanical loading had started to take off in NSW in the 1930s, and by 1939 accounted for almost 10% of coal produced. In 1945 the proportion increased to 22%, and by 1950 to almost 40%. The use of coal cutting machines spread rapidly in the South Maitland coalfield in the early 1900s, and by 1938 coal cutting machines were responsible for around 38% of the state's production. During the war little changed and the figure was 37% by 1945, but there were big differences across the coal districts, with the south almost 50%, and the west lagging with only 18.5%.

In the 1951-52 JCB annual report, we get a clearer picture of the state of mechanisation in NSW. The JCB categorised mines into four groups: in group A all coal produced was machine loaded and the mines had modern haulage systems; group B was all machine loaded, but without modern haulage; group C was partly machine loaded, and partly hand loaded; group D all hand loaded. Group A mines accounted for 20% of production, group B for almost 12%, group C almost 38% and group D almost 31%. So over two thirds of production that year still came from collieries in groups C and D, the two categories of mines which could reasonably be described as primitive operations.

Production in NSW in 1950 reached approximately 12.8 million tons, up almost 20% on the disastrous national strike-affected year of 1949. It surged again in 1951 and hit a record 15 million tons in 1952. Mechanisation was proceeding rapidly throughout the NSW industry, with the emphasis in the early 1950s on the mechanisation of the coal cutting and loading systems. The percentage of coal cut by machinery, only 40% in 1950, exceeded 51% by 1952, and over 63% by 1955. The percentage mechanically loaded followed a similar path, reaching almost 67% in 1955, compared with just under 40% in 1950. The JCB and the BHP/ AI&S group were the key forces behind these trends, although other companies were also moving ahead. By this time there were some mines in NSW that could fairly be described as state of the art operations, with modern machinery and equipment and amenities for the workers. A leading example was BHP's Burwood colliery, located near Dudley, south of Newcastle.

Burwood colliery was mechanised by BHP in the 1930s. By 1947, the colliery had 1280 kilometres of roadways underground, a reflection of its size and long life and 95% of the coal produced in 1947 was won mechanically. The colliery boasted coal cutting machines, power borers, locomotives for haulage, and a range of other machinery and equipment. Burwood even had an underground classroom for teaching mining trainees when they were not involved in the day to day mining operations. BHP also substantially mechanised its Lambton mine in the Newcastle district in the 1930s. And by 1950 it had effectively mechanised all its mines in the Newcastle district and its AI&S mines in the southern district. It purchased the Mt Kembla colliery in 1945 and immediately set about mechanising it. It also opened its new Nebo mine in the south in 1946 as a mechanised mine. Nebo also had a training school for new entrants to learn about mechanisation and mining operations.

In the four years to 1950-51, AI&S undertook an investment program arguably unprecedented in the industry up to that time (apart of course from the JCB's program). It purchased 45 locomotives for underground use, 30 mobile coal loading machines, 14 coal cutting machines and 240 coal wagons.[440] In this period the JCB bought more loaders than AI&S (46), but only purchased 6 locomotives; however its purchases also included 67 shuttle cars (used for coal loading underground) and 30 coal cutters. Except for shuttle cars, the coal producers including AI&S actually purchased more new machines than the JCB in this period. The JCB's purchases of machinery and equipment were for lease or sale to other producers, and also for use in its own collieries. By 1950 the BHP/ AI&S group was producing almost 2 million tonnes per year from an impressive stable of mines which included Lambton, Elrington, Burwood, Stockton Borehole in the north, Lithgow in the west, and Bulli, Wongawilli, Osborne Wallsend, and Nebo in south. By 1951, with production in excess of 2 million tonnes, BHP was finally able to satisfy the coal needs of its steelworks in Newcastle and Port Kembla from its own operations.[441] BHP was to continue to be the leader in the industry in terms of mechanisation in the 1950s, with its Newcastle district mines 84% mechanised by 1960 and its southern AI&S mines 96% mechanised (mechanised here defined as the percentage of coal produced by mechanical means).[442] By way of contrast, and as we saw in Chapter 5, in 1947 the big government-owned colliery in Lithgow, the State Coal Mine, which employed around 490 people and produced around 1650 tons per day, was still mining all of its coal by hand, and loading was also generally done by hand. There were no locomotives in the colliery, and haulage was still done by an endless rope system, a type of haulage that had existed in the industry for decades.

Another early 1950s example of a modern mechanised mine was Newstan colliery at Fassifern south of Newcastle. Newstan was developed by the JCB as part of its program to expand production through its own mines and was opened in 1950. Newstan was the renamed old Northumberland and Olstan collieries and became the JCB's model colliery, an example showing the way forward for underground mines. The JCB boasted in its 1949-50 annual report that Newstan had been "completely mechanised with the most modern equipment" and that it also had an excellent industrial relations record, having had no strikes (apart from the general strike in 1949) and also having a low rate of absenteeism.[443] In 1950 Newstan was still in the development phase, but expanded over the following few years.

In June 1950 the JCB announced that the first 3 continuous mining machines ordered from the Joy company in the USA had arrived in Australia, 2 for use by the JCB itself and 1 for BHP, and that it had chosen Newstan to test the machine in Australia. The continuous miner replaced four operations in an underground mine – drilling the coalface, shotfiring, cutting and loading. The JCB said that the machines were not suitable for use in all mines, being designed to operate in seams between around 1.5 metres to 2.5 metres thick. The continuous miners offered a number of benefits for the industry, including higher productivity, the elimination of shotfiring which was always potentially dangerous, and reduction in dust as the machines were fitted with water sprays. There were also critical safety benefits, as the men operating the machines were also located a distance away from the coal face under the roof which was supported by timbers, with the machine also having its own hydraulic jacks which could directly support the roof or help support the roof while timber props were put into place.[444] It would be several years before continuous miners became widely used in the industry, but they would spread rapidly throughout the underground sector later in the 1950s and in the 1960s, and would prove to be one of the major advances in underground coal mining in the post war years.

By 1953 Newstan had progressed to become the largest colliery in Australia, with production reported to be 2700 tons per day, overtaking BHP's Burwood which was then producing around 2400 tons per day. A glowing article in the Newcastle Morning Herald said that Newstan, which was still being developed, had been described by visiting mining experts as "the most modern and highly mechanised colliery in

Australia, and comparable to the best mines in America." It had only 300 employees, but "no mine can approach its record output of nine tons (per) employee".[445] Presumably this reference to productivity related only to the production per underground employee.

In early 1954, Newstan's brand new coal washery opened after testing. JCB chairman Sam Cochran said that Newstan was "the last word in mechanisation – as all mines will have to be if the coal industry is to live in the competitive coal age we are now entering…The mine is mechanised from the coal face to the railway wagons, and the average daily production at the colliery is 6.47 tons per man…The overall rate for the whole industry is three tons per man."[446] Cochran added that a number of other coal producers were introducing coal cleaning equipment, and that when those installations were complete, a substantial proportion of NSW coal would be mechanically cleaned, meaning a major improvement in the quality of the coal being produced and sold throughout the State.

Mechanisation proceeded during the 1950s, but standards of installation and maintenance of machinery were not high. The NSW Mines Department's chief inspector of coal mines noted in the 1959 annual report that, while further progress had been made in mechanising coal production, there had been no improvement in the poor standards of installation and maintenance of fixed machinery he had noted the previous year, in both the underground and open cut sectors. The Chief Inspector also noted that it was still "quite common to find the interior of a colliery workshop resembling a mediaeval blacksmith's shop."[447]

Industry mission looks at mechanisation overseas

JCB chairman Sam Cochran, NSW Colliery Proprietors' Association chairman Edward Warren, and Miners' Federation check inspector for the northern coalfield, Jack Barratt, travelled to the USA, Great Britain and Western Europe in 1952 to investigate mechanised mining operations. On the group's return, Warren said that the future of the industry in Australia would depend on the adoption of mechanised methods of mining like those employed in the USA. He said that mechanisation in the USA had not only been profitable for the producers, but also for the miners, as it had maintained or increased their rates of pay, and resulted in less physical effort. In relation to the mechanical removal of pillars, which was one of the key issues for the group to examine, Warren said that this process did not happen overnight, but had happened over a number of years, developing into a science that was well understood and accepted throughout the USA.[448]

Warren and his colleagues in the NSW Combined Colliery Proprietors' Association would also have been encouraged by the support for mechanisation from John L Lewis, the president of the United Mine Workers Association, the US equivalent of the Miners' Federation. Lewis was quoted in the NSW Colliery Proprietors' Association magazine in June 1952 saying that "I think mechanisation of coal gathering techniques is important for humanitarian reasons as well as economic reasons ….Mechanisation in the United States has robbed the industry of some of its elements of slave toil." In relation to his union's support for mechanisation over the previous 30 to 40 years, Lewis said that "… in return for encouraging modernisation, the utilisation of machinery and power…the union (had insisted) on a clear participation in the advantages of the machine and the improved techniques." Lewis also acknowledged that the coal mining industry in the USA still had some way to become more modern, efficient and profitable, saying that there were many mines which were not modern, with poor productivity, inadequate capital, poor quality coal and inefficient management. Lewis also claimed that the USA would have lost the war if its coal industry had not been able to supply the coal required for the massive war effort.[449] The US coal industry was certainly way ahead of Australia's at that time. In 1948 the NSW coal industry was producing only around 37% of coal which had been mined with coal cutting machines, compared with over 90% in the USA and 75% in the UK. The Queensland industry was even further behind, with the PDTS report in 1948 finding only 3 mines using coal cutting machines.

However the coal industry in both NSW and Queensland was changing and was set on a path to modernisation which would have many benefits and significant costs. The resistance by the Federation and its members to mechanisation would also ease, and other major changes including the widespread adoption of bonus systems for mine workers' pay incorporating a significant element tied to a mine's production would also occur.

Queensland Coal Board maps the way forward

The QCB, created in the wake of the establishment of the Joint Coal Board in NSW, was a major force for change in the industry in Queensland, but had a relatively small staff, more limited powers than the JCB, and the task of managing a very different industry structure. The QCB was created by the Coal Industry (Control) Act of 1948 which came into operation on 1 January 1949. The Members (directors) of the Board were Edward Dunne, previously the Under Secretary of the Department of Mines, Idris Evans, previously manager of a NSW underground coal mine, and Alfred Crowley, previously the chief electrical engineer with a UK colliery company. The creation of the QCB demonstrated that Queensland was determined to go its own way on coal industry regulation, although its coal industry was subject to the jurisdiction of the Coal Industry Tribunal which was created by the NSW and Commonwealth Coal Industry Acts of 1947.

The QCB's first annual report was in 1952, covering the period from the Board's inception to June 1952, and it gives an insight into the issues the industry faced. While Queensland had not experienced the same level of industrial turmoil as NSW, its coal industry nevertheless was in dire need of modernising and restructuring. Among many challenges facing the Board, the main one was how to increase production when the great majority of collieries were small operations with very limited financial resources.[450] In 1952, of the State's 83 underground collieries, 48 produced less than 100 tons per day (or less than around 25,000 tons per year). There were another 25 collieries producing less than 150 tons per day (or less than around 37,500 tons per year). Many of the mines were also poorly designed and would prove impossible to mechanise, even if the funds could be found for the necessary investment in machinery and equipment.

The only solution for the QCB was to ensure that these small collieries installed some basic equipment (and in particular power borers and pneumatic picks), eliminated haulage bottlenecks and reduced or eliminated hand wheeling, the movement of coal in skips by brute human force. [451] Eliminating hand wheeling had a number of benefits, including making workers available for other operations, and allowing wooden skips to be replaced by metal skips or wagons of greater capacity. Hand wheeling was also difficult or impossible if roadways were too steep, and there were other methods including the use of horses, or preferably mechanical haulage systems, which could cope with steeper grades. [452] The longer term solution of course would be for many of these small mines to close, and for new and larger mechanised mines, properly designed, for example with roadways wide enough to allow the efficient use of machines, to be developed. The roadways, tunnels and shafts in many of the existing small mines were simply too narrow to permit modern machinery to be installed and operated.

The QCB did make progress in its first few years and by 1952 was able to point to the installation of power borers in almost all collieries, the reduction of long wheeling distances underground, the partial elimination at some mines of hand wheeling and the substitution of horses or mechanical means, and improvements in ventilation through better mining methods at several mines. It also reported that year that it had ordered a number of mines to install better methods of coal handling on the surface which also provided reasonable protection for workers. All of these factors, it said, had improved the working conditions of mine workers. Recognising the shortages of many materials and manpower and the poor layout of many mines, the QCB said that some substantial progress had been made, and it believed that subject to the availability of materials and manpower further progress would be made in the future. [453]

Other major challenges in the early 1950s were the need to reverse the drastic decline in the quality of coal coming out of most collieries and the need to put in place a strong system to control the opening and closing of mines. For a number of years, coal quality in Queensland had been on the way down, with half of the State's mines selling only run-of-mine coal (ie the coal which was sold as it emerged from the mine), generally without any washing to reduce the proportion of dirt, rock and clay etc, or without screening to produce consistent sized lumps of coal for different customers. As an indication of the declining quality, in 1940 the average ash content of coal produced in the West Moreton was 16%; by the end of the war this had grown to around 22%; and by 1948 to over 25%. By 1952 the position in the West Moreton had stabilised, but the Board still had a major task ahead to encourage or force producers to install wash plants or take other steps to lift coal quality.

In 1952 the Queensland Act was amended to clarify the Board's power to control the opening and closing of mines. The Board pointed to past practices where mines had been developed without a clear understanding of the coal resources they contained, or without the owners having the financial resources to develop them efficiently, or where mines were abandoned and valuable coal reserves locked away. In May 1952 the Board issued an order requiring that any new mine or mine re-opening would require its approval, and approval would only be given if the owners could demonstrate that the mine had sufficient reserves and that they had the necessary technical and financial capacity to establish an operation to a certain standard, with appropriate mechanisation. The order also prevented the closure of any mine which still contained economically recoverable reserves.

Over the next few years, progress in mechanising mines was slow, but at least there was progress. In 1949 only 2 collieries had power boring machines, but by 1952 almost all the collieries in the West Moreton had power borers or pneumatic picks.[454] A number of collieries in the Ipswich area had installed coal cutting machines in the early 1900s, however by 1955, the QCB's annual report showed that only one coal cutting machine was in use in the state in a colliery in the Rosewood district. Another colliery had purchased a coal cutter, but found that it was unable to install it due to the poor power supply to the mine. By 1955 there was also only one mechanical loader in use in the underground industry, located in one of the mines in the state's northern region. However there was more progress with mechanising haulage, with 7 diesel locomotives installed in collieries in 1954-55.[455]

After the Powell Duffryn report in 1949 the coal producers were involved in a dispute with the Government on the question of who would pay for the mechanisation of mines. In 1950 the producers had seen the agreement between the Victorian and Queensland Governments for the Commonwealth to subsidise the supply of 900,000 tons of Callide coal to Victoria over three years. This was a great win for Callide, but raised concerns in the West Moreton about what support there might be for that region's producers. That year the producers proposed to the Queensland Government that it follow the example of the Joint Coal Board and establish a pool of machinery and equipment which would then be leased to individual producers. One major producer said that the cost of some items of machinery was very expensive and beyond the reach of the majority of producers. The Government rejected that idea, but instead undertook to provide finance at a concessional rate 3.5% to allow producers to buy machinery.[456] A Government loan of $250,000 to the Queensland Coal Board was then approved to assist the Board to purchase machinery and equipment and to assist the development of collieries. In announcing the loan, the Mines Minister said that some producers had already applied to the QCB for finance to purchase equipment.

It is also not generally understood that loan funds were also made available from the World Bank to Australia in the early 1950s. The first loan for $100 million was advanced in 1950 and was used for a variety of purposes including development of electric power generation, water conservation projects, railways, agriculture, mining, smelting and refining, steel production and a range of other manufacturing industries. The Bank's second loan to Australia was for $50 million to be used for financing development of a number of sectors. For coal mining the intention was to finance the importing of equipment for both open cut and underground mining operations. Four collieries in the West Moreton benefited from this program.[457]

Pillar extraction breakthrough achieved

For most of the industry's history in Australia, underground mining accounted for the bulk of coal produced. However for many years a high proportion of coal in underground mines was never mined, but left as pillars to support the roof of each mine. A proportion of coal in pillars was extracted by hand, but this was far less than could have been produced by mechanical means. Hand mining of pillars was also was also slow due to union restrictions on the number of miners allowed to work on each pillar. In the Maitland and Cessnock area, in most mines only one or two pairs of miners were allowed to work, and with each pillar holding between 4,000 and 9,000 tons, it took up to 2 years for a pillar to be extracted. But there was a major problem with the heating of the coal which meant that in practice a high proportion of what could have been easily mined coal was never mined, as sections of mines had to be sealed to prevent heating.[458] The NSW Government-appointed committee which investigated the problems on the South Maitland coalfield in the early 1950s looked at 8 major mines and found that over half of the pillar coal was left in sections of those mines which had been sealed off due to heating.

Removal of these coal pillars by mechanical means was common in other countries, but had been banned by NSW Mines Minister Jack Baddeley in 1941 with the support of the Miners' Federation. In June 1948 the JCB called in experts from the US Bureau of Mines and the British Coal Board to advise it on mine mechanisation, including mechanical extraction of pillars, and the effects of subsidence. At that time in the average mine about half the coal was left unmined in the form of pillars. The Miners' Federation position was that it was dangerous to extract pillars, while the NSW Colliery Proprietors' Association position was that mechanical extraction was normal in every advanced coal-mining country, and that it would open up millions of tonnes of coal lost to the nation because of the policies of the NSW government and the Miners' Federation.[459] The US and British experts reported to the JCB in August 1948. They concluded that extraction of coal could be accomplished successfully using the best mining practice, and that this would lead to better safety, more efficient operations and reduced loss of coal.[460] The experts found that about 170 million tons out of the total of 239 million tons of pillar coal could be extracted safely. The report also said that wherever pillars could be extracted safely by hand, they could be extracted even more safely by mechanical means, and that quick extraction was the fundamental basis for successful operation and greater safety. The Miners' Federation however was not convinced. Its general president, Idris Williams, said that the Federation would differ sharply and violently from aspects of the report, and that it "opposed the view that recoverable coal in pillars should be extracted as quickly as possible and that as "some of the pillars had been standing for about 100 years… their extraction would be extremely dangerous."[461]

The potentially mineable coal in pillars represented over ten times the industry's annual production in NSW in 1948. Access to such a massive resource was to be a major issue for the industry for some years. Despite the experts' report, implacable opposition from the Miners' Federation meant that the old hand extraction system would continue for several more years. The Federation's arguments regarding safety included their claim that the noise from machines would prevent miners from hearing the sound of movements in the roof of the mine, these sounds being an indication of the possible collapse of the roof or part of it. The Federation was also concerned that the space required to operate and manoeuvre machines would make it difficult to install sufficient timber props to hold up the roof; it also had concerns about dust produced by machines and the effect on health.

Another problem was that the producers were reluctant to invest in expensive new machinery for working at the coal face and for haulage if the same equipment could not be used for extracting pillars.[462] Many of the industry's engineers and proprietors argued that it was neither practical nor economic to install expensive equipment to mine coal from the bords and cut-throughs if they were denied the right to use the same equipment to extract the pillars. As the JCB noted, it was for this reason that some colliery owners stated in no uncertain terms that they were not prepared to risk major investments in mechanisation until the issue of the mechanical extraction of pillars was resolved. The JCB also pointed out that even if the pillar issue was resolved, it would take some time to increase production, as mines' equipment was already busy and new equipment would have to be ordered and installed, and haulage systems would also need to be reorganised.[463]

In 1949, the producers had a potential breakthrough when the Minister for Mines gave Wallerawang Colliery near Lithgow the go-ahead to extract pillars mechanically, although the Minister's approval was subject to advice from a committee on the most suitable method of mining. The committee was to include representatives of producers, employees, the JCB and the Department of Mines. The JCB's annual report for 1949-50 stated that the employees had failed to nominate their representative and the committee had been unable to function.[464] The Miners' Federation wanted no part of the committee's work and was able to continue to block any progress. In June 1949 the NSWCCPA applied to the CIT to amend the coal award to provide for mechanical extraction of pillars. The CIT brought down its decision in April 1950 agreeing that the producers in NSW, Queensland and Victoria should be allowed to mine pillars by machinery, but this decision was rendered irrelevant by the Federation's opposition. The JCB was clearly in favour of mechanical extraction and referred to Davidson's conclusion in 1946 that the alleged dangers which were the basis for the Federation's objections were without foundation and were "mere propaganda directed to delaying progress in mechanisation."[465] The JCB queried the Federation's motives, saying that it was "...quite evident ...that the Federation had decided that the question of the mechanical extraction of pillars should not be decided on its merits but would be retained as a bargaining point."

Following the return of the overseas mission led by the JCB chairman in 1952, NSWCCPA chairman Edward Warren and Federation check inspector Jack Barrett were supportive of mechanical extraction. Warren said that he "saw no reason why mechanisation of both solid and pillar coal should not be introduced in Australia for the benefit of everybody in the industry." Barrett said that, provided suitable conditions of ventilation and transport of both men and coal were introduced, he was prepared to give favourable consideration to the proposals of the producers. However the Federation was still adamant, although its position had softened and no longer was based on outright opposition. Despite having a district check inspector as part of the mission, the Federation said that it was not officially represented on that mission and had no say on which countries or mines were to be visited.[466] The Federation maintained that it "would not depart from its opposition to the mechanical extraction of pillar coal under present conditions" and that it was "timely to recall that an undertaking was made that stowage would begin in (NSW) at the beginning of (1952). We are still waiting." The Federation's central council's formal position was that mechanical extraction was unacceptable until the industry had demonstrated its sincerity on stowage.[467]   Stowage was the term used to describe the process of filling in the area left after the pillars had been removed; the filling could be done with rock or other material and once completed would prevent the roof from collapsing. However this was a costly exercise and required large quantities of inert materials to be obtained and transported underground.

The producers operating on the Greta seam were particularly keen to move ahead with mechanical extraction of pillars. The Greta seam was prone to spontaneous combustion (spon com) and stowage would assist to reduce the potential for spon com by blocking off the air flow to areas already mined. While the Federation's major stated objection to mechanical extraction was the risks to miners' safety, employers and the JCB believed that the real objective was to delay the threat to jobs which were seen as the inevitable outcome of most forms of mechanisation. In 1951, the South Maitland Coal Conservation Committee (with representatives from The Department of Mines, JCB, producers and the union, and chaired by a university expert) recommended that a stowage scheme be implemented on the Greta seam mines, financed by a levy on all underground mines. This was rejected, but in February 1952, the JCB and several producers agreed to proceed with stowage trials, and funds were made available from the JCB. Five stowage projects were subsequently approved to proceed, but over the next 10 years, little progress was made towards finding a viable solution.

The Federation's opposition however could not delay the issue for ever, and in August 1954 the CIT gave all mine workers an increase in pay where mechanical loading was in place. Following this decision, the Federation put the issue to a vote, with a recommendation that mechanical extraction of pillars should be accepted. The members voted to accept mechanical extraction of pillars, but the great majority did not vote and the vote in favour was not overwhelming. It appears that many mine workers had accepted the inevitable and saw a vote as a foregone conclusion. In the NSW northern district, of the approximately 1500 who voted (only around 15% of the Federation's district membership), the vote in favour was only around 53%. The vote in favour in the south and west was also quite narrow, but again with only a minority of members attending mass meetings and voting.[468] The vote in Queensland in September saw a much more decisive mandate with over 80% of those voting agreeing to the recommendation. The stage was now set for the producers to access the valuable pillar coal, and there was a steady increase in pillar production in the next few years. The lifting of the Federation ban also saw a major increase in investment by the underground producers in new machinery and equipment including coal cutting and loading machines to extract the pillars. This investment of course was not just a result of the lifting of the ban, but also due to more attractive coal prices and improved profitability.

There were potentially many millions of tons which could now be extracted from pillars, but there were also practical problems to confront, including the fact that as many pillars had been mined years earlier, they had deteriorated in quality and were not suitable to mine. However "extraction began with whatever equipment was available" with continuous miners used extensively as they were well suited to the task. And once mines had begun to mine the readily accessible pillars, they started to develop underground workings which were designed for pillar extraction, with "layouts that permitted pillars to be taken as soon as possible after formation."[469] In June 1955, just over 5% of underground production in NSW was won from pillar extraction, and by June 1957 the proportion had grown to 17.5%. The figure for the 1959-60 year was almost 30% and this increased to 45% by 1964-65.[470] Clearly, major volumes of coal were being mined which would have previously been left in situ, but 10 years after the restrictions were lifted by the Federation, the proportion unmined was still significant.

Industrial relations improve

In the pre-war period, the NSW coal industry came under the jurisdiction of the Commonwealth Arbitration Court for issues that affected more than one State, and the NSW Industrial Commission for intrastate matters. Under its wartime powers, the Commonwealth set up the Central Reference Board (CRB) and Local Reference Boards in 1941. In 1943 the CRB was replaced by the Central Coal Authority (CCA), with A C Willis as the Chairman, with its jurisdiction restricted to issues relating to the Miners' Federation. The CRB continued as the tribunal with responsibility for matters affecting the other craft and staff unions. In 1944 the CCA became the Central Industrial Authority, and the Local Reference Boards became Local Industrial Authorities.

These arrangements continued until the Coal Industry Tribunal commenced operating in March 1947. Under the Coal Industry Acts, the jurisdiction of the CIT was restricted to Miners' Federation matters, as was the case under the wartime arrangements with various tribunals. The Joint Coal Board stated that the wartime arrangements were continued as the craft unions were afraid that if they were forced to come under the same jurisdiction as the Federation they would lose their separate identity and character, "particularly in view of the fact that industrial unionism (ie one union for all classifications in the industry) had long been the official policy of the Miners' Federation."[471] The Acts also set up Local Coal Authorities under the CIT, responsible for disputes involving the Federation at the mine or local level. Francis (Frank) Gallagher was appointed in February 1947 as the head of the CIT. Gallagher was only 42, and had spent the previous 15 years working for the NSW Railways, mainly on industrial issues. Gallagher would stay as head of the CIT for the next 28 years.

With the establishment of the CIT and LCAs, disputes and other matters such as awards involving the craft unions continued to be handled by the Central Reference Board and the Local Reference Boards. The dual system of the CIT/ LCA and CRB/ LRB continued for several years. Gallagher was also appointed Chairman of the CRB to look after craft union matters and so became responsible for industrial issues for the industry in two quite separate jurisdictions, one under the Coal Industry Acts, and the other under wartime security regulations which continued after the war had finished. The dual system involving the CIT and the CRB was finally rationalised in December 1951 when the NSW and Commonwealth Governments amended the Acts to bring the craft unions under the jurisdiction of the CIT.

The 1950s saw a dramatic reduction in industrial disputes in NSW. In 1949-50, 14.2% of shifts were lost to industrial disputes, an average of around 36 shifts per year for every employee in the industry. This was fortunately the peak, and the loss in 1950-51 dropped to 7.1%, after which the trend was down, with only 4% lost in 1958-59, and only 2% lost in 1959-60. The 1960s would also see a generally lower level of disputation, until the trend was reversed in the early 1970s. The NSW coal industry, employing around 19,000 workers in 1950, accounted for 47% of all days lost in the State. By 1959, with the coal industry's disputation down, the industry's share of total days lost was down to 29%. A relatively small industry in terms of employment was still the major source of days lost in all industries.

JCB and CIT have a positive impact

It is difficult to apportion credit for the decline in disputation in the 1950s which was arguably the result of many factors. The work of the JCB and CIT, although often criticised by the producers and the unions, gave the industry a more stable base on which to operate, and so appears to have been beneficial. The Newcastle Morning Herald certainly gave the JCB and CIT its approval in November 1952, when conditions in the industry were looking up. The NMH editorialised that "The new prosperity of the industry, in which employees are sharing through improved conditions, greater security and unprecedentedly continuous work, is basically the result of the increased demand for coal since 1939. But it would not have been possible without the supervising, coordinating and directing influence of the Joint Coal Board, the support of State and Federal Governments, and the constructive efforts of the Coal Industry Tribunal and its local agencies. It has been a demonstration of the effectiveness of team work. One of the greatest accomplishments of this effort has been the sublimation of the bitterness in employer-employee relationships that used to be the bane of the industry."[472] The NMH went on to say that "The emergence of the Coal Board as a buffer between the companies and the employees has been an important factor in bringing about this better spirit." However with the emerging problem in late 1952 of a surplus of coal, the newspaper cautioned that "the bad old days are too recent and the memory of them still too fresh in the minds of many to permit the assumption that they could not return, or that the time has come to risk their return."

However the story for the decade was not as simple as the JCB and CIT steadying the ship. The mechanisation of the industry and the gradual disappearance of the contract system were also important factors, as was the spread of incentive payment schemes which began to be adopted in some mines in the mid-1950s and had become more common by 1960. The JCB's annual report for 1959-60 noted that "These schemes under which over-award payments related to mine output are paid to all employees have given the employees concerned a direct interest in increasing the productivity of their mine."[473] These schemes continued to spread rapidly through the industry in the 1960s. Another factor may also have been the so-called "attendance bonus" that the CIT awarded in December 1950. This decision gave mine workers an additional day's pay for those who had worked a full ten-day fortnight. This was an interesting innovation by the CIT, and was initially rejected by the Federation and the mine workers, but was soon accepted as one of the award conditions. Of course shifts were also lost to a range of other factors including absenteeism, workers compensation and sickness. In NSW in 1949-50, industrial disputes accounted for 60% of all shifts lost, the total loss rate being over 23%. That overall rate of shifts lost had dropped to 9.5% by 1960, a major improvement, but still unacceptably high.

In Queensland the rate of disputation had been traditionally lower than in NSW, perhaps a reflection of the structure of the industry, with many mines owned by small family operated companies which had closer working relationships between the mine workers and the managers and owners. In 1949, the year of the national strike, industrial disputes were the cause of 8.2% of shifts lost in Queensland. The overall rate of shifts lost however was 17%, with absenteeism the other major cause. Apart from 1951, when 4.5% of shifts were lost to disputes, the figure in Queensland was never above 3% in the 1950s, and had dropped to only 0.7 % by 1960. The overall rate of shifts lost in Queensland fell from its peak in 1949, but remained stubbornly high in the 1950s, with the 1959 figure 10% and the 1960 figure 9.3%, almost the same level as in NSW.

Impact of Incentive schemes and mechanisation also positive

In December 1950, the Coal Industry Tribunal made an award which granted mine workers payment for an additional shift on the condition that they worked at least 10 days in the fortnight. In announcing his decision, Gallagher said that in his opinion the "coal industry will become stable when, and only when, consistent and regular attendance at work is obtained from the employees...I think that an order which might secure this attendance is requisite."[474]

It would have been unthinkable for this to happen in almost any other industry at that time, and most workers in Australia would have jumped at the chance to earn an extra day's pay just for coming to work every normal working day. But this was not a normal industry and the Miners' Federation objected to the award saying that the allowance constituted an incentive payment which was contrary to union policy, the longstanding contract payment system notwithstanding. Idris Williams, general president of the Federation, said that the Federation would not be bound by a decision that would not bring standards applying to industry generally to mine workers. "The award provides for incentive payments on the basis of attendance by mine workers. The industry is already overloaded with incentive payments, such as the contract system. Additional incentive payments can only bring about industrial dissatisfaction."[475] This time the Federation decided to take a new approach and in January 1951, rather than closing all the collieries by striking for a period of time, the Federation leaders recommended to its members stop work for one day per week. The recommendation was approved.

Following the CIT decision, the Queensland Coal Board adjusted selling prices to cover the cost of the CIT decision. The secretary of the Coal Owners' Association, unhappy with what it saw as an inadequate increase in prices, said that the producers might take drastic action to protest against the QCB's decision by closing down their mines or moving to transfer control of the industry to the Federal Government.[476] With Sydney in particular suffering major cuts in electricity and power rationing, the Federation's action was roundly condemned. The Newcastle Morning Herald, which was generally more restrained than papers such as the Sydney Morning Herald in its criticism of striking miners (its full title being the Newcastle Morning Herald and Miners' Advocate), this time took the Federation to task. On the day that mine workers were to vote on the Federation's recommendation, the NMH questioned the sincerity of the union, saying it had tolerated the payment of rewards and incentives to its members for many years. What the CIT was offering, the paper said, was a system of payment that many of its members were already enjoying, adding that the colliery proprietors had been happy to pay for the continuity of regular work.[477] The NMH also appealed to the mine workers to recognise the harm that they would cause to other workers and to the supply of gas and electricity which was critically dependent on the supply of coal. The Federation's members accepted its leaders' recommendation and the one day a week strike action proceeded.

In early March the Commonwealth Government announced urgent plans to introduce legislation to amend the Coal Industry Act to give the CIT the power to enforce its awards and containing provisions for heavy fines and imprisonment for those who disobeyed the wards. However with the legislation also requiring a similar amendment of the NSW

Coal Industry Act, the Commonwealth Bill could go nowhere. NSW Premier McGirr put the issue before his Cabinet which concluded that amendments to the Act were necessary, but that the details needed to be fine-tuned by the NSW and Commonwealth legal experts.[478]

The NSW Government's non-decision prevented the Commonwealth from proceeding, but the proposed amendment to the Acts were overtaken by events including an application by the producers for the award to be varied to provide a temporary lifting of the weekly hiring provision, a change which would have put extra pressure on the Federation by giving the producers the power to stand down workers at any time during the week. The JCB also became involved, expressing its concern that strikes were slowly strangling the Australian economy, and urging the CIT to call a compulsory conference to bring the producers and the union together. Gallagher accepted the JCB's recommendation and the conference went ahead.[479] Gallagher rejected the producers' application and also criticised them for opposing an application by the Federation for a one week adjournment of CIT hearings to wait on the results of mass meetings. Gallagher referred to the producers' "high and mighty" attitude, and accused them of showing a "complete disregard for the welfare of the community." Gallagher struck out the producers' application and adjourned the hearings for one week, during which workers accepted the Federation recommendation that they end the one day stoppages. Gallagher then agreed to review the award, drawing the ire of the producers.

Gallagher's comments and his decision saw the NSW Colliery Proprietors' Association chairman, Edward Warren, come out fighting. Warren said that the producers who had been criticised by the CIT were the same producers who had consistently obeyed the Tribunal's orders and had not been allowed to run their own businesses. He went on to say that it was difficult to believe how Gallagher could accept the Federation's refusal to accept his December decision while taking this attitude to the producers.[480]

However Gallagher then issued an order in early May that the award would stand, incurring the wrath of the Federation. The ACTU president

Albert Monk, who had become involved in the dispute, warned the Federation to accept the decision and encouraged the miners to cease their strike action. JCB took the unusual step of writing to all mine workers with a copy of the CIT's order. The JCB's letter said that it wished to explain the CIT decision to mine workers so that any misunderstanding was avoided. One interpretation of the JCB's action is that it feared that the Federation leadership may not have been able to be trusted to convey the CIT decision faithfully. The JCB's letter said that it appeared that there had been a good deal of misunderstanding by employees regarding the CIT's decision in December. It said that the attendance allowance was not an incentive payment as "opponents" claimed, and that under the CIT's order mine workers were not required to work any harder or longer than normal. "If you are one of the many mineworkers who customarily work and produce coal on 10 days a fortnight, you are getting an extra shift each fortnight merely for what you are doing at the moment" the letter said.[481] The letter also said that the order did not conflict with genuine union principles, as demonstrated by the fact that the original idea of the additional payment (11 shifts paid for 10 worked) had come from the board of management of the union's northern branch. The JCB also argued that the order would assist to avoid the "many petty unauthorised stoppages" which plagued the industry, stoppages which the union leadership had on many occasions condemned.

The strike action began to peter out after the CIT order was made and by late May it was apparent that most mines in the State were working continuously. In fact in the fortnight ending 18 May an estimated 85% of mines had worked a full 10 days, entitling their employees to the allowance. A number of mines however were still holding out, with the workers still to be convinced to cease the strike action. The end of the dispute effectively occurred when mass meetings of northern district workers accepted a recommendation from the union to protest against bonus payments, but not to take any direct action against the award. "The effect of this decision was to accept the bonus payments, but to refuse to allow it to interfere with the federation's traditional rights. These rights included a right to hold aggregate meetings and one day stoppages when considered justified."[482] It was reported that the northern district meetings were poorly attended, perhaps indicating that the Federation's members were taking a more pragmatic attitude to Gallagher's decision than some of the senior officials of the union. The union's principles were maintained, and life went on. In November 1952, JABAS chairman Thomas Armstrong said that his company's experience was that the attendance bonus had reduced the number of wildcat strikes, particularly towards the end of each fortnightly pay period, as workers were reluctant to lose a day's pay. However he also said that the downside was that there had been a tendency for strikes to last for several days once the bonus had been lost in any fortnight.[483]

Despite this significant dispute at the start of the decade, the years ahead would prove to be much more settled in terms of relations between management and workers, although of course, still very problematic. There were many reasons for the improvement in the industry's record, including rising wages and better conditions for mine workers, incentive schemes, the spread of mechanisation and the consequent abolition of the contract system, and the role of the JCB as an organisation which in effect stood between the workers and management, but which was able to achieve major changes in the industry.

The spread of incentive schemes in the industry during the 1950s and 1960s was arguably one of the most powerful change agents the industry saw in the post war period. Incentive schemes which tied workers' wages to production began to appear in the mid-1950s, with S&M Fox's South Clifton and Burragorang Valley mines the first to introduce them.[484] Unlike the old contract system, where the hewer and some other categories of workers employed on a daily basis had their pay linked directly to their own production, the new schemes were based on the mine's total production and so embraced all the workers at the colliery. All the workers in a colliery now had the opportunity to earn a production bonus and so had a direct financial interest in the production and the efficiency of that colliery.

The Miners' Federation leaders however did not take kindly to the initial introduction of these schemes at the Fox mines. Bill Parkinson, the general president, and Bob Cram, the southern district president, tried to persuade the miners not to agree to these schemes on the basis of their fears about safety. Parkinson said that on one occasion when he was standing on a 44 gallon drum addressing the members, the reception was so fierce that he feared for his safety and he had to jump off the drum, and both he and Cram had to scurry away. Parkinson also said that while the Federation's leaders continued to campaign against incentive schemes, their acceptance by the rank and file was so widespread by the early 1960s that the union had to accept the schemes as a fait accompli. Parkinson told one historian that the feeling of the members was so intense that continued opposition by the union leadership would have led to the destruction of the union.[485] The Federation's fears about safety concerns, whether they were genuinely held or used as a pretext to argue against change, did have some foundation. We will come back to the issue of production bonuses and safety later.

In 1958 the JCB noted that the recent introduction of these schemes had resulted in a jump in production from the mines involved.[486] By 1959-60, the JCB was reporting that 17% of NSW coal mine workers were employed on the basis of an incentive scheme, with their average earnings around 25% above workers in the industry not be paid under these schemes. The 1960s would see an even more rapid spread of these schemes.

Mechanisation was another key factor in the downward trend in disputes. As mines mechanised, the old contract system became obsolete, and the many problems associated with that system disappeared. Of course mechanisation had also been a major cause of disputes since the mid-1930s when BHP began the industry's inexorable push to modernise. In a submission to the CIT in 1960, Federation general president Bill Parkinson stated that mechanisation had eliminated a number of industrial disputes in the industry and had led to a complete revolution in mining production methods.[487] Referring to the South Maitland coalfield Parkinson said that in 1951 unmechanised mines had produced 96% of coal, but by 1959 that figure had fallen to only 11%.

Quite apart from the contract system, the impact of mechanisation on industrial disputes was difficult to assess. The JCB annual report for 1959-60 contained some interesting data on the disputation record of the five categories of collieries which were grouped according to their degree of mechanisation. The state's underground mines as a whole lost 4.9% of shifts to industrial disputes in 1954-55, and by 1959-60 the loss had fallen to 2%. However there were significant variations in the proportion of shifts lost by the different categories of collieries, with the lowest loss rate recorded by the non-mechanised mines employing less than 12 workers underground.[488] This group of mines however accounted for only around 1% of production in 1959-60 (down from around 18% in 1954-55), and so were likely to have been small family owned operations. In 1954-55 the same category of mines had lost 5% of shifts, or around the overall industry average. A category of mines which were partly mechanised (category C – partly machine loaded, partly hand loaded) had a loss of only 0.6% of shifts. The category with the worst record in 1959-60 was group D, (all hand loaded, and employing more than 12 workers underground), which lost almost 3% of shifts.

Queensland mechanisation proceeds slowly

By the end of the 1950s, the Queensland industry had made some progress towards fully mechanised operations, but still had a fair way to travel. The industry was still dominated by small collieries, with 33 out of a State total of 76 collieries producing less than 100 tons per day, and another 20 producing less than 150 tons per day. There were some underground operations of reasonable size, including the Collinsville State 1 and 2 mines which produced around 156,000 tons in 1959.60, and Bowen Consolidated which had a similar output. In the West Moreton district, the largest underground colliery was Box Flat Extended which produced just under 200,000 tons in that year. The open cut sector was still in its infancy: Thiess' Kianga operation produced only around 37,000 tons in 1959.60, and its Callide open cut around 68,000 tons. At Blair Athol, there were two open cut mines which produced a combined output of around 168,000 tons.

By 1956, there was still only one privately owned mine in the state which was using a coal cutting machine and face conveyors. However interest from the owners and managers of collieries was also reported in several districts in mechanisation of the mining operations at the coal face. By the mid-1950s, the use of diesel locomotives underground was growing, with 30 locomotives in use in 18 of the state's mines. And ventilation was improving as old inefficient fans were also giving way to new efficient axial flow fans.[489] In 1959-60, Box Flat Colliery near Ipswich became the first Queensland mine to install a continuous miner, together with Joy loaders and shuttle cars. That year Box Flat also ordered a second continuous miner, and the early indications pointed to these machines being well suited to the colliery's geology and conditions. Tivoli Haighmoor colliery had also placed orders in 1959-60 for a continuous miner, loader and shuttle cars and several other collieries had also moved to adopt modern machinery and equipment including loaders and conveyors.

Mechanisation was of course not a simple process of ordering new equipment and installing it. Companies needed access to finance, skilled engineers and tradespeople, proper mine planning systems and the appropriate mine layout and design to cater for modern machinery and equipment. The Queensland industry in the 1950s was lacking in all of these areas. And the shortage of mine surveyors (a role which was critical in mines undergoing mechanisation) was a particular problem. There was also a need for skilled people to plan for and manage the installation and maintenance of machinery, and there was a distinct lack of what the QCB called "forward planned development" at mines which were looking to mechanise. In its annual report for 1959-60, the QCB reported that it was common that "various items of plant have arrived at a mine, yet cannot be put into operation because (a) the mine development is not complete to receive them, or (b) some vital ancillary (equipment) such as electric cables or control units has not been ordered in sufficient time."[490] The QCB went on to point out that it was not a sensible use of capital to purchase expensive items of plant and then have them lying idle for long periods of time. It reported that some collieries had purchased equipment more than a year earlier and were still to install it.

The level of mechanisation in Queensland was quantified more clearly for the first time in that QCB annual report for 1959-60, which contained a summary of the production and output per shift for the first half of 1960. For the six month period, the Queensland underground sector produced 1.1 million tons of coal, with only 154,000 tons (or just under 14%) produced by mines which were categorised as completely mechanised. A further 239,000 tons (21%) of production came from partly mechanised mines. The fully mechanised group achieved an output per manshift (OMS) of around 5.3 tons, compared with around 2.9 tons for the partly mechanised group and for the unmechanised mines employing more than 15 men underground. The benefits of mechanisation were becoming evident, but the industry clearly had a long way to go before it could be described as fully mechanised.[491]

Queensland emerged from World War Two with a major coal quality problem – the average ash content its coal produced was around 25% in 1948, up from 16% in 1940.[492] The Powell Duffryn report had found that in 1949 only six Collieries in Queensland had a washery, with five of these in the West Moreton, and with one other colliery in that district in the process of constructing a new washery. All of the washeries were found to be inefficient and primitive. For a state looking to build new power stations to supply electricity to a growing population and industrial base, poor quality coal was no longer acceptable, and the QCB had as one of its priorities the need to encourage collieries to invest in plants which could wash coal to remove much of the impurities and supply a higher quality product to power stations and other industries.

However progress was slow and it was not until 1956-57 that the QCB was able to report that a feature of that year was positive action by some of the collieries in the West Moreton in providing "modern and satisfactory" coal cleaning and beneficiation plants. That year saw one colliery actually having installed a washery, with three other collieries in the process of constructing plants and two more intending to order new plants.[493] The QCB said that the investments in washeries by these six collieries, which accounted for 38% of West Moreton production, would have considerable benefits for coal consumers.

It would be the 1960s and 1970s when we would see a dramatic transformation in the industry in Queensland as massive new mines are developed in the Bowen Basin and many of the small mines, particularly in the West Moreton, cease to operate. The 1960s would see the great majority of mines turn their backs on the old hand mining methods and adopt modern machinery and equipment.

Australia forced to import coal

From the late 1940s, with the Industry in NSW still rocked by disputes, shortages of coal forced the Commonwealth to subsidise imported coal. Victorian and South Australia were particularly hard hit by the shortages, as they had traditionally relied on NSW as a major supplier. These States had imported small quantities of coal during the 1949 national strike, but with NSW still unable to meet the demand, Victoria and South Australia ordered 3 million tons of coal from South Africa and India. By mid-1952 around 1.2 million tons of this imported coal had been delivered. These coal imports were subsidised by the Commonwealth, the subsidies averaging around $5 per ton. The supply of coal from Thiess' Callide mine to Victoria was also subsidised.

Disputes were not the sole problem with restricting the supply of NSW coal to other states. National Development Minister Spooner said in July 1952 that it was a pity that there was substantial coal already stockpiled in the NSW northern and western districts, with transport problems preventing this coal being delivered to customers.[494] However the fact that Australia had to turn to overseas suppliers to meet its coal needs was an indictment on the industry. Despite the best efforts of the Joint Coal Board, coal production in NSW was not able to be increased fast enough to satisfy local demand. The JCB's program in its early years was based on its estimates that demand for NSW coal would increase rapidly, from less than 15 million tons in 1949 to 18 million by 1953. Without the major disruptions in 1948 and 1949 the JCB said that the industry would have produced around 13.6 million tons in 1948-49. With disputes knocking out production of 1.5 to 2 million tons a year, the JCB said that production would need to grow by around 50% over the following 5 years.[495]

While BHP was pushing ahead with its own development program, the JCB was focussed on forcing other companies to mechanise and developing its own open cut and underground mines. In its first year, the JCB had surveyed all the coal producers to determine their plans for the development of their collieries. "The replies indicated that with a few exceptions no plans for future development had been prepared and that in the majority of cases, none were contemplated. In some cases this was due to lack of capital; in others to lack of appreciation of the necessity for such development; and in others to the attitude that it was useless investing capital in the rehabilitation and mechanisation of mines without industrial stability."[496] The JCB had also realised early on that, because of the woeful industrial disputation record of the industry, the NSW coal companies would struggle to raise the necessary capital to finance its much needed modernisation and expansion.[497] The JCB therefore developed a range of strategies to tackle the problem and quickly increase production. In its early years a major priority for the JCB was the development of its own new mines in the three districts, and the assumption of control of some mines on a temporary basis. It recognised that development of any new underground mines would take at least 3 years, and so its early developments included a major program to open up new open cut mines, including undertaking exploration and drilling work to obtain detailed understanding of the coal reserves. When the JCB commenced to operate in 1947, the known reserves of coal which could be mined by open cut methods were a paltry 5 million tons.[498] Other major strategies included the establishment of an equipment pool, with the JCB purchasing a range of machinery and equipment to be leased to the producers or made available for sale. A second strategy related to pricing, whereby the JCB would set prices for producers which would allow them to make reasonable profits and so generate some sorely needed capital to fund the purchase of machinery.

By 1950, over 50 collieries had accessed the JCB's equipment pool, with around half of the machinery and equipment leased to the collieries and the rest purchased outright. As time went on and profitability improved and companies' access to finance improved, the proportion of equipment purchased increased. Included in the JCB's early equipment purchases were two continuous miners manufactured by Joy in the USA; one of these was allocated to the JCB's own Newstan colliery on a trial basis, the other was purchased by AI&S for one of its Illawarra collieries.

In the 1949-50 year, the JCB's own production division produced almost 900,000 tons, a worthy achievement for such a new organisation. Production from the JCB's mines reached 2.3 million tons in 1951-52, or around 16% of the total NSW production, placing the JCB as the number two producer just behind the combined BHP/AI&S group. But 1952 saw the shortage of coal give way to a surplus, and also saw the outlook for strong growth prove too optimistic. The industry in NSW was now producing too much coal, particularly from its open cut mines. The JCB submitted a plan to the NSW and Commonwealth governments for the stabilisation of the industry which included a levy on coal producers to be paid into a stabilisation fund. The fund proposed would have had three major objectives – to promote the export of NSW coal, to compensate open cut producers for contracts which the JCB may need to break, and to provide financial assistance to displaced coal workers.[499] The plan was rejected by the Commonwealth and by the producers who were strongly opposed to the idea of a levy on production, which they saw as just a cost impost which would harm their competitive position. The Commonwealth, no doubt influenced by the attitude of the producers, also tried to block the JCB from becoming involved in the export market. NSW Mines Minister Arthur said that the Commonwealth had told the JCB that it was not its function to meddle in the distribution of coal for export.[500] However cooperation to promote export was on the horizon and the producers and the JCB would soon form an export committee to look for overseas markets and promote the State's industry, a committee that the Commonwealth would also join in due course. In December 1952, in response to the over-production in the industry, the Commonwealth

Government announced a series of measures to support production in NSW, including the financing of stockpiles and provision of finance for sales of coal to interstate customers.

In 1951, the JCB had forecast that demand for NSW coal would increase to over 17 million tons in 1952, and to 19 million tons in 1954. These forecasts proved to be way too optimistic. By the time of the publication of its annual report for 1951-52 in August 1952, the JCB had slashed its forecasts, with the 1952 demand expected to be only just over 14 million tons; the Board's 1954 demand forecast was reduced to just under 16 million tons, with only marginal growth in 1955 and 1956. The market outlook had fundamentally changed and the JCB now had to severely cut back its own mining operations, with its open cuts to be the target in order to protect employment in the underground sector. The JCB had developed 19 open cut mines by mid-1952, but as the new market situation became clear, it moved quickly to close many of these mines; by the end of 1953 it operated only 11 open cuts, and by 1956 only 1. It would later also extract itself from underground mining.

Within just a few years, the JCB was also able to take credit for the best overall industrial relations record for any of the major underground producers in the industry. For the 1952-53 year, its underground operations lost just over 8% of shifts to industrial disputes, compared with almost 16% lost by the industry in total. The NSW Government's own mines, operated by the State Coal Mines Control Authority (and including the big Lithgow State Coal Mine) were only marginally below the industry average, although somewhat better than the large private sector producers including BHP, AI&S, JABAS and Caledonian.[501]

Coal prices, subsidies, taxes and profits

In the 1930s, coal producers struggled to make any profits, with many producers losing money, running down their equity in their businesses and failing to invest in new equipment. BHP and some other producers including JABAS of course did invest in expansion and modernisation in the 1930s, but with the outbreak of World War Two the industry was poorly placed to meet the challenges ahead. From early 1943, when the Commonwealth's price stabilisation program took effect, coal prices were fixed and from that date it was necessary to provide subsidies to compensate producers for rises in wages and other costs. When the JCB commenced operating in 1947, it inherited the price structure which had existed before the war, although prices had been adjusted for those costs increases. The JCB described the price structure at that time as 'extremely uneven and often illogical.'[502] It also saw that the subsidy system had failed to give the producers a financial incentive to increase production and improve their efficiency, and it decided to do away with the subsidy system as soon as it was practicable to do so; this was a policy which the Commonwealth had been pursuing. The JCB also noted that another drawback of the subsidy scheme was that producers were compensated for the costs of industrial disputes which lasted up to 4 days. From April 1947, in one of its first major decisions, the JCB increased coal prices in all three districts, and at the same time reduced the subsidy to producers from around $6 million per year to less than $2 million.

The JCB also understood that because of the industry's poor industrial record, it was difficult for companies to raise funds to finance investment to mechanise their mines. Price increases were a feature of the JCB's policies in its early years, and helped the industry in NSW to start to earn some consistent profits. This in turn helped to change the mindset of the industry from generally negative to more positive, and to one which was prepared to look at investing for the future. Early in 1952 the JCB introduced a new price fixing policy designed specifically to allow producers to earn reasonable profits and to provide a more attractive environment for finance to flow into the industry. That same year also saw an important change to the Income Tax Act to put the coal industry on the same basis as other mining. For many years coal mining companies had been unable to claim expenditure on mining plant, development, housing and welfare buildings as a deduction against their taxable income. Changes to the Act were passed in December 1951 and took effect in the 1951-52 financial year.

The mood in the coal industry had certainly shifted by 1952. However a serious new problem now emerged – coal production was no longer unable to meet the demand; in fact a serious over-supply had developed. The new market situation was caused by several factors, including changes in interstate markets, with domestic supply in Victoria and South Australia now able to meet the local needs of power stations. Victoria's brown coal industry had been on an expansion path since the 1920s and the state was now self-sufficient. South Australia's Leigh Creek mine was in production and supplying coal to the Port Augusta power station. The price increases granted over the previous years to NSW coal producers, while restoring some needed profitability and confidence, also meant that NSW coal had become relatively expensive. In fact the JCB stated that coal prices were now too high, with local Victorian and South Australian coals much cheaper per ton than NSW coal delivered to those states, even though the energy value of NSW coals was higher. The early 1950s also saw the start of the challenge from oil as a major competitor. Oil was cheap and coal had become too expensive, and of course coal's reputation for unreliable supply courtesy of the traditional industrial warfare in the industry had not endeared it to customers in factories who were now looking to adopt fuel oil to power boilers and other equipment. During the years when coal was in short supply some consumers had switched to oil; more were now looking to do the same.

The JCB itself had believed that many coal users who had gone to oil would switch back to coal, but admitted in its 1952-53 annual report that this had not happened.[503] Fuel oil had traditionally been imported into Australia, but there were new oil refineries planned which would produce petroleum products for cars and other uses, but also residual products which could substitute for coal and be sold as a cheap by-product to industrial users. And if those trends were not enough, another of the coal industry's major traditional markets, railways, were now introducing diesel powered locomotives. The costs of operating diesel locomotives in the early 1950s were around 45% less than for coal fired locomotives. Since the 1800s, the railways' steam locomotives had been one of the coal industry's major domestic consumers, but the days of the steam locomotive were numbered. In 1953 Australian workers including coal miners, won increases in the basic wages in May and August. The JCB's new policy now was to force the producers to absorb the cost of these wage increases.[504] The producers now had to confront declining local markets and also absorb higher costs.

Menzies and McEwen open trade relations with Japan

In 1951, only six years after the end of the war, and with feelings in Australia towards its recent enemy still raw, Prime Minister Menzies signed a peace treaty with Japan. This was followed shortly after by goodwill visits to Japan by Menzies and his Foreign Minister, Percy Spender.

While John 'Black Jack' McEwen was the driving force in the process which would lead to a formal trade agreement with Japan, Menzies himself was also involved in selling the case for closer trade relations to the public. As former Prime Minister John Howard notes in *The Menzies Era*, Menzies was arguing the case for greater trade with Japan as early as 1953 in regular radio broadcasts, emphasising the imbalance in trade, with Australia's wool exports far exceeding our imports of goods from Japan.[505] McEwen was Minister for Commerce and Agriculture from 1949 to 1956, when his functions were divided between new Ministries of Trade (which he then held) and Primary Industry (which was allocated to William McMahon). McEwen took over as Leader of the Country Party in 1958, at which time he also became Deputy Prime Minister. In 1954, the Menzies Government took the first concrete step towards a closer relationship with Japan when Cabinet approved a start to informal discussions with the Japanese on trade relations. By 1956 McEwen was in a position to recommend that negotiations proceed to a formal trade agreement, and the Australia Japan Treaty on Commerce was signed by McEwen and Japanese Prime Minister Kishi in July 1957. Despite the tensions so soon after the war, Menzies was magnanimous towards the Japanese, saying "We have not, in spite of what may have been the problems of ten years ago, approached the problems with Japan in a spirit of hatred or unpleasantness. On the contrary, we have made our watchword, full and friendly association."[506]

The Treaty was bitterly opposed by many at the time, with the Labor Opposition under Dr H V (Doc) Evatt opposing it in Parliament. The major concern of the Opposition, unions and manufacturing industry groups was the potential for imports of Japanese goods, particularly textiles and clothing, to damage our local industry. The Victorian Chamber of Manufacturers was particularly scathing saying: "On 6th July, Australia's manufacturers received at the hands of the federal Government one of the most crushing blows ever dealt to them..." [507]

It was also in 1957 that the Treaty of Rome was signed; this was the agreement between the major Western European powers that created the European Economic Community, and in due course, the European Union. The European countries which had created the European Coal and Steel Community in 1951, had now taken the next step, and while the UK held out joining the EEC initially (and its entry was blocked for a time by France's President de Gaulle), it would only be a matter of time before it became a member. De Gaulle lost power in 1969 and the UK went ahead with another application to join the EEC. In January 1973, the UK, Denmark and Ireland joined the original EEC members in the new and enlarged union. UK membership of the EEC was critically important at the time for Australia as, up until that time, we had enjoyed special access to the UK market for our agricultural exports. Our agricultural exports to the UK were now severely affected, but by 1973 the minerals industry in Australia was beginning to boom, heralding a new era for the country. As Howard notes: "If the Commerce Agreement had not been concluded, it is doubtful whether the transition from Britain to Japan as our best customer could have been so readily achieved."[508]

From the Australian Government's viewpoint, the 1957 Treaty was largely focussed on the potential for higher exports of agricultural products including wheat, wool and sugar. The potential for Japanese imports of coal and other minerals was not really on the Government's radar at that time. However the Treaty was certainly timely and can only have eased the way for the massive boost to our minerals industries which was soon to occur as a result of the post war growth in Japan.

Australian coal industry grows while Japan's falters

To understand the Japanese interest in Australia in the late 1950s as a potential coal supplier to Japan, it may help to look briefly at the history of the coal industry in Japan. The Meiji Restoration of 1868, when the Japanese imperial system was restored under Emperor Meiji, began a period of development which resulted in Japan entering the twentieth century as a modern industrial nation, with a strong industrial base and a powerful navy and army. In fact Japan was powerful enough to overwhelmingly defeat Russia in the war of 1904-05, the first major international conflict of the century. This period also saw Japan's coal industry develop into the country's major source of energy, and the coal industry was to continue to grow until the post-World War Two era. By the early 1900s, Japan was also a significant coal exporter and competitor with Australia for Pacific markets. Its exports in 1903 totalled around 2.5 million tons, including around 0.5 million tons for use in ships. Japan's principal markets were China, the USA and the Straits Settlements (British colonies which are now mostly part of Malaysia and Singapore).[509]

As Japan prepared for conflict and developed its military strength in the decade leading up to World War Two its coal production more than doubled. And at the peak of its war efforts, coal accounted for almost 60% of Japan's total energy needs. During the war, Japan's coal industry was a vital cog in its economy and production averaged around 55 million tonnes per year, or 4 times the average output of Australia's industry. At its peak, Japan's wartime coal industry employed almost 400,000 workers, many of whom however were prisoners from China and Korea; and several thousand prisoners of war (including some Australians) were also part of that workforce.[510] Following the end of World War Two, initially under the umbrella of the US occupation, it did not take long for Japan to begin to once again transform itself into an industrial powerhouse, and the growth of this Asian tiger would have a profound impact on Australia.

But in understanding the rapid growth in the post war Australian coal industry, it is also important to understand the dramatic changes which shook the Japanese coal mining industry from the late 1950s. In many ways, the decline in the Japanese coal industry and the transformation and growth of the industry in Australia in the post war years were opposite sides of the same coin. In the year commencing April 1945, during which Japan was occupied by the Allied forces, crude steel production had fallen to around 0.5 million tonnes. However by 1950 production had recovered to over 5 million tonnes, and by 1960 had soared to around 23 million tonnes. By 1960 Japan's steel industry was the country's number one export earner and Japan had moved to number three producer in the world behind the USA and USSR. The Japanese economy was booming and needed growing supplies of quality coking coal to feed the demand for steel from Japan's domestic and export industries and for steel to export to Japan's growing foreign markets.

But a major part of the story of Japan's hunger for imported coal to feed its steel industry, and later its electric power industry, is what happened to its own coal industry. From the early 1950s, rationalisation began in Japan's coal mines, with for example Mitsui sacking almost 3500 workers in 1953, but reinstating around 1800 of these workers following pressure from the coal mining union.[511] In the 1950s Japan had a "coal first – oil second" energy policy which looked to an expanding domestic coal industry to continue to underpin its economic growth. However following the end of the Korean War, Japan's economy and its coal mining industry suffered, and in response the Government implemented its Coal Mining Temporary Measures law in 1955, which set up a special corporation to manage the rationalisation of the industry. The new program established targets for the industry, including production levels, efficiency and costs. The objective was to raise production from around 43 million tons in 1954 to 51 million tons in 1960, while at the same time lowering coal costs and therefore prices through higher productivity which would be achieved partly by closing inefficient mines.

By around the time of the Suez crisis in 1956, demand for coal in Japan was again growing, and coal mining companies were expanding, taking advantage of government subsidies, and unions were forcing up wage rates.[512] Restrictions on the importing of oil and coal were still in place, shielding the coal industry from competition. In 1957 a new strategy was adopted, still based on coal first – oil second, and again aiming to expand production. The following year MITI was forecasting that production would reach 69 million tons by 1967, and that reductions in employee numbers, if necessary at all, could be achieved through natural attrition.[513] But opinions in Japan on the future of coal and the coal first policy were divided. At a major industry conference at the end of 1956, for example, it was argued that the high price of coal was a factor that was holding back the country's economic development, and that Japan needed to convert to oil as its principal source of energy.[514] By the late 1950s, it was becoming apparent that its policies were putting unreasonably high hopes on coal as the primary source of energy for Japan's future supply of energy."[515] There was significant pressure on the coal mining companies to improve their efficiency, with the Government issuing a dictate to companies in 1959 to reduce their production by 20% on 1957 levels. The companies responded with cutbacks, their industry association saying that they would need to reduce their overall employment levels by 100,000 by 1963.[516]

The response of many companies was to ask their employees to agree to voluntary retirement programs. The coal unions resisted, but by around mid-year, most disputes between the unions and companies were settled. The major exception however was Mitsui, which had announced a program of 5000 voluntary retirements. The union refused to cooperate in identifying workers to participate in the downsizing. The company then dismissed almost 1300 workers and in January an indefinite strike began. The strike lasted over 300 days and ended in 1960 in defeat for the miners. That strike and its outcome heralded the start of a rapid downward spiral for the industry.[517] The Japanese Government however did move to assist the coal mining regions which were severely hit by the cutbacks in the industry by funding projects for unemployed mine workers and coal communities. In 1961, the industry federation, Keidanren, announced that its steel and electric power company members had agreed to contract for between 13 and 20 million tons annually until 1967, thereby giving the coal industry a base on which it could plan. Oil imports were freed up, and MITI put a halt to exploration for any new domestic coal deposits. But in 1962 we see another change of plans. A task force set up by the Government proposed that coal production should increase to 55 million tons a year by 1967 by encouraging the development of new power stations and increasing subsidies for coal production. The Government accepted these recommendations, which were passed into law in 1963.

The maintenance of a significant domestic coal industry in Japan was not made any easier by some horrific disasters in some of its mines. In 1963 Mitsui's Miike mine was again in the news, but this time for all the worst reasons, when an explosion killed 458 workers. In 1965, an explosion at the Yamano mine in Kyushu took 236 lives. Nevertheless, the Japanese Government was determined to maintain domestic coal production, and changes to coal policies continued to be made. In 1965 the Government implemented the recommendations of another review of the industry to cut the production target to 52 million tons, with an increase in prices and also subsidies for mining companies to invest in capital equipment. However the pressure from oil continued and further reviews of the industry led to continued downward revisions to production targets and subsidies for companies to close mines.[518] These subsidies had a major impact, with many mine closures occurring. During the period from 1946 to 1972, subsidies to coal companies totalled around 260 billion yen.[519] The Japanese coal industry reached its post war peak in production in 1961 of over 55 million tons, but by 1973 that figure was down to around 21 million tons. Employment in 1960 was around 244,000, but by 1973 had dropped to around 25,000. There were 682 mines in 1960, but only 57 by 1973.[520]

The 1950s and 1960s saw a confusing series of policies and policy switches, but in retrospect, the writing was on the wall for the coal industry by the late 1950s. The massive cut backs in employment and mine closures devastated many communities in Japan, but coal was a high cost energy source, oil was cheap and plentiful, and market forces were not to be denied. Over this period, oil imports into Japan soared, so that by the time of the first oil crisis in 1973, oil accounted for 75% of Japan's energy supply compared with only around 20% in 1950. The oil shocks in 1973 and in 1979-80 would bring home to Japan the fact that it had become dangerously dependent on oil for its electricity generation and general industrial needs. But that dependence was to prove a boon for Australia from the late 1970s when Japan decided to change its energy mix once again and become a major importer of steaming coal.

It is no surprise that the major Japanese coal mining companies, including Mitsui and Mitsubishi, were prominent in the push to develop the Australian industry from the 1960s and 1970s. These massive companies were also Japan's traditional link to the world as trading companies and benefited from their role as middlemen in the coal trade that developed between the two countries. But coal was also in their DNA: Mitsui purchased the Miike coal mine on Kyushu from the Government in around 1889; Mitsubishi's entry into coal mining came a few years earlier in 1881 when it purchased the Takashima mine on an island near the major port city of Nagasaki. Mitsui and Mitsubishi would continue to play a major role in the Australian coal industry and the export coal trade and the Japanese coal industry would continue to stagger along well into the 1990s, suffering the death of a thousand cuts.

Japan steel companies now look to Australian coal

The Australian coal industry emerged from World War Two as an insignificant exporter. From a peak of almost 3 million tons in 1912, Australian coal exports fell during the World War One, rose again after the war, and peaked again in 1922 at around 2.4 million tons. From there it was downhill, and by 1947, exports had fallen to only 46,000 tons. The focus of the industry over the early post war years was to meet the demands of the domestic market, but it was in 1952 that the industry, the Joint Coal Board and the Federal Government began to look seriously at the potential for export markets to be developed. By 1952, the efforts of the JCB and the producers had lifted NSW production to 15 million tons, or around 50% above the 1945 level. There was potential for growth in the demand for coal from the expanding Australian steel industry, but BHP and its subsidiary AI&S had that market sewn up. Other domestic markets were not looking promising, and NSW was actually losing markets in other States as their own coal production grew. Exports from Australia were still negligible, with only 233,000 tons of coal exported from NSW in 1952, 69,000 tons going to Asian countries and 164,000 tons to Pacific Islands.

Australia's interest in export markets for our coal began to get serious in September 1952, when a NSW industry mission, led by JCB Chairman Sam Cochran, and including Edward Warren (chairman of the NSW Colliery Proprietors' Association and CEO of JABAS) visited a number of Asian countries. The JCB established an Export Committee in January 1953, with representatives from the Board and the NSWCCPA, and in May the Commonwealth Government also became involved, with its Director of Trade Promotion joining the committee. That committee met regularly over the next couple of years. January 1953 also saw four ships loaded with coal to Japan; there was only a very modest quantity (33,000 tons) involved and these shipments did not herald the start of a resurgence in the export market.

In evidence to a case before the CIT in November 1953, the JCB's Secretary said that the Board had responded to orders from a number of countries, including Egypt, Italy, Pakistan, Burma, Indonesia, Hong Kong, The Philippines, Chile and New Zealand, but the only order it obtained was for 100,000 tons to Korea. In its evidence to the CIT in 1953 the JCB said that it had found that the international market was extremely competitive, with "Australian supply and quality far above that which customers were willing to pay", and that it was not worth the trouble investigating other overseas markets.[521] Despite its quality, our coal was clearly not competitive in many overseas markets. The major focus of those early efforts to secure export markets was on steaming coal for power generation and coal for the production of gas. Coking coal for steel production was not yet high on the agenda.

In 1954, what was billed as the first Australian coal mission left to investigate prospects in Korea, Singapore, The Philippines, Japan and Burma. Led by A E Warburton, a Member of the JCB, it included the managing director of Muswellbrook Coal (representing the NSW producers), and a fuel technologist. National Development Minister

Senator Spooner released the group's report on their trip in December 1954, the report saying that Japan was interested in the possibility of buying coking coal, but that there were no reasonable prospects for the sale of steaming coal to other Asian markets. While Japan was stated as having a surplus of gas coal, the group found reasonable prospects for gas coal in Hong King and Singapore.[522] However in retrospect, that mission was a year or two too early; a mission a little later in the decade would have focussed on coking coal and by that time Japan was starting to look to diversify its sources of supply.

In the early 1950s, Japan was relying on a mix of local and imported coking coal for its steel industry. Imports were around 3 million tons a year, and the USA dominated the import market, with several other countries, including the USSR, China and India supplying the balance. In 1954, with the assistance of the Australian Embassy in Tokyo, NSW coal producers offered to supply 500,000 tons of coal to Japan. In response to this proposal Mitsui sent a representative, Mr I Kitahara, to Australia to assess the coal industry and its capacity to supply. Kitahara recommended that samples of coal be sent to Japan for testing. This was done, but the tests proved unfavourable, and little Australian coal was exported to Japan in the subsequent few years.[523] In 1956 a trade mission from Japan to the USA concluded that the country in future needed to look to suppliers located closer to it, rather than on suppliers located on the US east coast.[524] A coal strike in the USA that year also gave a ray of hope for the industry in NSW when some Japanese steel mills begin to look to Australia as another supplier of coking coal.[525]

A little known company, Blits Trading Company, received a request in 1956 for the supply of 50,000 tons of coal from Newdell, a Hunter Valley mine then owned by the JCB. The Japanese request led to a firm order from Fuji Iron and Steel and Yawata Iron and Steel. Shortly after, trading companies Heine Bros and Gollin & Co received orders, Heine from Kawasaki Steel Corporation through Marubeni, and Gollin an order from Fuji through Mitsui. The orders were soon met with exports out of the port of Newcastle. Further orders were also negotiated. These were small orders, and clearly the Japanese steel mills were buying to test the quality of NSW coals and the industry's ability to export.[526] Jaques Blits had migrated from Belgium to Australia and built up a commodity trading business. Blits went on to develop an ongoing export trading business with Howard Smith, mainly selling coal from Howard Smith's Caledonian Collieries. When Caledonian was merged with JABAS to form Coal & Allied Industries Ltd in 1960, Blits declined an offer to run the export business of the new company. However his staff joined Coal and Allied and became the nucleus of its subsidiary company Coal & Allied (Sales).[527]

Export prospects for Australian coal began to improve in 1958, when the first senior group of Japanese coal buyers arrived in August to look over the coal industry in NSW and Queensland. The month before, JCB Chairman Cochran had said that the Board believed that there was potential for Australia to export around 2 million tonnes a year to Asia. American hard coking coals were regarded as top quality and were what Japan needed to blend with its own soft coking coals to produce quality steel. But the US supplies had to be shipped from the east coast of the country, incurring high sea freight costs. Australia had the potential to supply coal at a lower landed cost, given the shorter distance to ship the coal, but the Japanese would need to be convinced that we could be competitive on a cost basis, have acceptable quality coals, develop the necessary transport infrastructure and be a reliable supplier.

The big breakthrough then came in December 1958, following the Japanese mission, when the Japanese steel mills signed a contract for 1.2 million tons over 5 years with NSW south coast mine Coalcliff, followed only a week later by a contract for 720,000 tons of coal with another NSW South Coast colliery, Corrimal, also over the next 5 years. These orders were followed by a small order for 30,000 tons from Burragorang Valley colliery Nattai-Bulli, and another order for coal from S&M Fox's Wollondilly colliery. The Nattai-Bulli and Wollondilly orders followed visits to Japan by the CEOs of both companies. By late 1958, the JSM were also negotiating a long term contract of 1.3 million tons with South Clifton colliery, also located on the NSW south coast; this contract was confirmed in February 1960. These contracts with NSW coking coal suppliers marked the first significant move by the Japanese steel mills into the Australian coal industry. Following years of restructuring and turmoil in the NSW coal mining industry, the 1958 contracts gave real hope that the industry's prospects had turned the corner. However, while the NSW industry had made some strides towards becoming more efficient, the JSM put the industry on notice that it would need to lift its game. This became clear in early 1959 when the Japanese Iron and Steel Federation, the industry organisation representing the steel producers, said that Australia needed to cut the cost of both mining and transporting its coal.[528] This marked the start of international pressures on the Australian coal industry and its infrastructure providers to lift their game – to improve efficiency, lower costs, upgrade rail and port facilities. The market potential was there to capture, but it would not be won without effort.

Japanese steel producers had established links with the coking coal producers in the USA on whom they relied for most of their imported coking coal requirements. They had begun to understand the potential of Australia as a major supplier, but also knew that the Australian coal industry had a long way to go to reach a reasonable level of efficiency. The Japanese also knew that Australia's rail and port facilities would need substantial upgrading and expansion to cater for any significant lift in exports. The Japanese Iron and Steel Federation said that "Generally speaking, the mining efficiency (in Australia) is low at two or three tones per miner but efforts are being made to raise it to about ten tons per day through modernisation…If this succeeds in reducing the mining prices from the present ($US11-12 per ton) to …. ($US5-6), these coal fields will become a permanent source of supply of coking coal to Japan over many years." The Federation said southern fields near Port Kembla were the most promising. [529]

While the prospects for the NSW coking coal producers had certainly lifted, the next few years would see the start of the development of the Queensland coking coal industry, and the competition between the NSW and Queensland producers would be fierce.

The coal crisis - the decline of the Greta seam mines

While the export prospects for the Australian coal industry were looking more positive by the late 1950s, and the domestic market for steaming coal was looking brighter as the NSW Government began to plan for new electricity generation, the industry in the Cessnock district in NSW was going through a crisis. In the 1950s, a number of the major Greta seam mines which had been developed in the early 1900s were still operating; some had been mechanised, while others were still only partly mechanised. By the mid to late 1950s, the outlook for these mines was becoming gloomier. Competition from oil to fuel industrial boilers intensified, the market for gas coal began to decline, interstate coal markets also began to decline, and the railways continued to invest in diesel locomotives and to phase out steam powered locomotives. The JCB in its annual report for 1955-56 said that that year had been the most difficult year that the NSW coal industry had had to face since the war. New oil refineries were now in production, the Snowy River Scheme was beginning to supply electricity into NSW and Victoria, and coal mining was seeing a significant increase in efficiency as a result of the mechanisation of mines. Consumption of coal in NSW fell by almost half a million tonnes, the first fall for many years, and employment in the industry fell by around 1500 or over 12%.[530] In the following year, 12 collieries closed, mainly due to the loss of markets.

In January 1956, the Hetton Bellbird Colliery Company announced the closure of the Bellbird mine near Cessnock, a mine employing around 600 workers and working the Greta seam. The year also saw the closure or cessation of work at a number of other mines in the northern district and the western district, including Millfield and Millfield North on the Greta seam. The shut-downs included some old established mines in the north working the Borehole or other seams, and included Stockrington No. 1 and West Wallsend Extended. In 1957 there were more mine closures, including Stanford Main No.1 (originally named Stanford Merthyr colliery which was developed from 1902) and Lambton Central No.1. The Miners' Federation called the situation a crisis for the coal industry and in particular for the Cessnock and surrounding areas.

In March 1956, following news about Stockrington and closures on the
western coalfield totalling another 400 jobs, the Federation's general
secretary George Neilly said that the shutdown of mines in northern New
South Wales could lead to the biggest industrial crisis in the coal industry
since the Depression. Neilly warned the State and Federal Governments
that this position could not continue, and asked the President of the
ACTU to contact State and Federal Governments to secure the jobs of
mine workers. At the ACTU Congress the previous year it was decided
unanimously to support the miners in efforts to retain their jobs.[531]

Later that month the Federation decided it was time to abolish the
JCB. Federation general president Bill Parkinson said that the JCB had
been established to bring about economic security to mine workers.
"It has not been following that policy, but has been carrying out the
policy of the Federal Government" Parkinson said. This has resulted in
the "retrenchment of hundreds of men from the industry. We say that
the Joint Coal Board has ceased to play a useful role."[532] Parkinson said
that the Federation planned to ask the Commonwealth Government to
transfer its powers to the NSW Mines Department. The Federation saw
this step as clearing the way for the NSW Government to implement the
Federation's policy of nationalising the industry.

A delegation comprising leaders of the Federation and representatives
of the Federal and NSW Labour caucuses and the ACTU secured a
meeting in March with the Federal Minister for National Development,
Senator Spooner, and the Minister for Labour, Harold Holt. Around
1,000 striking mine workers, their wives and members of allied unions
travelled to Canberra in support of the Federation, and staged a march
on Parliament House led by a number of the miners' wives. The allied
union members included members of the Waterside Workers Federation,
and the unions representing coal trimmers, shipping clerks and painters
and dockers (the Newcastle branches of these unions having gone on
strike in support of the Miners Federation). The delegation asked the
Federal Government to impose an excise duty on imported oil to make
it less competitive with coal. Not surprisingly, no agreement was reached
at the meeting. However one of those present, J D Kenny, representing

the ACTU, reported on the meeting to the assembled mine workers, wives and allied union members and said that the deputation had asked Spooner and Holt to carefully consider "proper control" of the industry and to undertake a detailed investigation into the utilisation of coal. The delegation also said that they would ensure that the NSW Government carried out its responsibilities.

It had been a tough year for the local industry, and the outlook was not improved in November when JCB Chairman Sam Cochran told a meeting of the Coal Industry Committee (a committee set up by the Federal Government to look at ways to assist the coal industry) that mines in the Cessnock area were over producing by around 2500 tonnes a week, and that in January 1957 theses mines would lose contracts of 4000 tonnes per week with the Victorian Gas Corporation. This loss would take the area's over production to around 6500 tonnes per week, and there were no new markets or contracts in sight.[533]

The closure of Bellbird did not eventuate until July 26, 1957. At the start of the last morning shift on the 25th, 36 miners began a "stay-down strike" and were joined by another 12 miners from the night shift. With dismissal notices for all 350 Bellbird mine workers becoming effective on the 26th, the striking miners said that they intended to stay down in the mine until the State of Federal Government or coal authorities found jobs for all the dismissed workers. For the previous 20 years, the Bellbird mine had been selling all its production to AGL, but by 1956-57, AGL was refusing to buy more from Bellbird as it could obtain cheaper coal from other collieries in the area. JCB chairman Sam Cochran was dismissive of the stay-down miners, saying that "The fact that only a few Bellbird men are in the mine indicates that the majority realise that the colliery is a victim of a failure to find a cost formula which would allow the mine to continue."[534]

The stay-down miners were supported by around 1800 other district mine workers from the Aberdare group of mines who went on strike on the 26th in sympathy with the Bellbird miners and also to protest against dismissals from the Aberdare West mine. The chairman of Hetton Bellbird Coal Company said that it was impossible to keep the colliery operating, although he hoped that he would be able to announce that the mine would re-open in the following couple of weeks. The stay-down ended on 7 August when the miners were told that efforts to convince AGL to purchase more Bellbird coal had been unsuccessful. The reopening of the mine did occur, but not until 1958. The mine kept operating, eventually closing in 1976. Bellbird hit the news again in 1965, but for a very different reason. On July 20 a skeleton and a watch were found in an old area of the mine; they belonged to George Bailey, the miner who had died in the explosion of 1923, but whose body had until that time not been found. Bailey had worked as a telephone attendant, and on the day of the explosions had not received a warning of the impending danger.[535]

The Miners' Federation pressed the NSW Government to address the unemployment crisis in the industry through measures including nationalising coal mines and reducing the working week to 35 hours. The Federation however was rebuffed by the Government, with Mines Minister Nott saying that these measures could not be justified, and with the State producing 20,000 tons per week in excess of demand, the cost of nationalising mines would be enormous. After meeting the Minister, Federation leader Bill Parkinson said that the Minister had pointed out that his Government had the power to nationalise mines, but not the resources, while the Federal Government had the resources but not the power. Parkinson's conclusion from the meeting was that coal mines would never be nationalised.[536]

As an indication of the concern about the future in the Cessnock area, a protest meeting was held in Cessnock in September 1958, with over 3000 miners and other local residents attending to hear speeches from politicians and union leaders. This was believed to be the biggest community meeting held in the northern coalfields since the national coal strike of 1949. All the mine workers in the district took the day off.[537] Bill Parkinson told the meeting that his union's membership had dropped from 24,000 in 1950 to 16,300 in 1958. A high proportion of those who had left the industry had worked in the Cessnock and broader Newcastle districts. The meeting passed a motion which called for the nationalisation of the coal mining industry, a shorter working week, construction of a plant on the coalfields to process by-products from coal, and protection against imported fuels. The meeting also asked the NSW Government to immediately begin construction of a local power station. This meeting did not produce any major results, but the nationalisation agenda of the Miners' Federation was not dead, and the planning for the new power stations in NSW was underway.

Sadly, the markets for the Greta seam coal were not expanding, and with mechanisation, the mines that served these markets were continuing to reduce their workforces, while other mines were closing. For the northern district overall, the decade between 1950 and 1960 saw coal mining employment fall by around 5,000, from just under 13,000 to around 7800. Employment in the district had held reasonably steady until 1956, but each subsequent year saw significant falls which were to continue into the late 1960s when growing export demand started to have a real impact and demand from the Electricity Commission and the steel industry also played a major role. The Cessnock municipality, which contained most of the collieries which closed or suffered severe cutbacks, was the area which was hit hardest by the restructuring in the industry. The municipality's population, which had declined in the 1930s, fell by around 3500 between the census years of 1954 and 1961.

One of the factors in the decline in the Greta seam mines was the policy of the major domestic customer AGL. AGL, the company which commenced in the 1830s and supplied the gas for Sydney's first street lamps in 1841, had a contract with PACCAL (Petroleum and Chemical Corporation of Australia Limited) to supply refinery gas to AGL's Mortlake plant on the Parramatta River from January 1956. The contract was for 20 years for the supply of residual gas from PACCAL's oil cracking operations and provided a major boost to AGL's supply capacity. However the price of the gas was "computed so that it would always be less than the cost of gas manufactured at Mortlake." This contract was also important for AGL because it "freed the Company from the constant anxiety that had resulted from its heavy dependence on northern coal."[538] Mortlake was the AGL facility which processed coal into gas. The market for Greta seam coal took a further blow when in December 1961 a second pair of naptha cracking plants came on stream at Mortlake, ending the supply of coal to AGL. This also saw the end of the sixty milers on the Parramatta River, the coal ships which operated between Newcastle and Sydney.[539]

Queensland underground job losses and quotas

In Queensland the industry was also losing jobs in the late 1950s. After employment peaked in the underground sector in 1955 at around 3500, around 400 jobs were lost over the next 5 years as some mines closed and others retrenched workers. The last mine at Mt Mulligan which employed around 70 men finally ceased operating in late 1957. In 1957 and 1958 the job losses led to a number of strikes and demonstrations in Ipswich and Brisbane, and in February 1958 a series of stay-down strikes took place as miners protested about cut backs. At the Caledonian Colliery at Rosewood, 70 mine workers staged a stay down in February 1958, protesting against the retrenchment of 24 men.[540] This action followed the cancellation of contracts with the South Brisbane Gas Company and the Railways, and was followed a few days later by a district strike involving around 2400 mine workers and mass meetings to consider reductions in coal quotas allocated to collieries by the QCB and retrenchments. The Mines Minister convened a meeting the day after the strike with the industry parties and asked the QCB to reconsider the quota issue. On the 19th the Minister announced that tentative orders had been arranged to keep Caledonian operating, and he said that he hoped there would be an early settlement of the dispute and that the workers at Caledonian would be re-employed. Work did resume at the colliery on the 24th following the restoration of a contract for 50 tons per day, and all the retrenched workers were re-employed except one who had found another job.[541] The Caledonian dispute was thus settled without too much pain, unlike many of those in the NSW Cessnock area. The Queensland underground sector, while suffering some significant job cuts, was able to continue operating without major disruptions into the 1960s, when mechanisation of most collieries took place. Underground employment in Queensland would fall to around 1500 by the end of the 1960s.

During the 1950s, one of the demands on the JCB was for it to impose quotas on production in NSW collieries. However the JCB firmly rejected this approach, arguing that "The Board has consistently rejected suggestions that the level of production should be controlled by the imposition of quotas on output from some or all collieries. The Board is convinced that a policy of quota control would raise more problems than it would possibly solve."[542] It stressed that quotas on production "… of themselves provide no assurance that the quota can be sold or that colliery revenues will be sufficient to continue employing any given number of men. In fact by reducing production, quotas might well impair the ability of a colliery to continue earning sufficient revenue to cover costs. A colliery in such a position would not remain open for long…… coal mining industry in NSW can make its most effective contribution… only if it remains efficient and … is able to respond quickly and effectively to changes in economic circumstances …Change is an essential feature of an efficient dynamic industry." [543]

The JCB however was not totally opposed to quotas, as there were what it called "consumer quotas", which did apply to the major customers for the coal industry. It argued that these quotas assisted changes in production to be gradual and "To this end the Board has sought to have consumer quotas determined from time to time by the large consumers, notably the Steelworks, the Electricity Commission and the Railways…" The JCB said it appreciated the willingness of these major customers to accept such a policy.

In 1962, in a publication which reviewed its activities since it was established, the JCB contrasted the pressures for change in the NSW coal industry compared with Queensland. It referred to "the relatively stagnant position which has characterised the Queensland coal industry during the last 10 or 12 years" and said that this "stability" was to a significant extent due to the fact that the market forces in Queensland were weak compared with the forces operating in New South Wales. "The operation of the quotas in Queensland may well have assisted in damping down any changes within the industry." The JCB however recognised that major changes were underway in Queensland by the early 1960s and that the industry there would have to undergo major structural change, with or without quotas.[544]

In the 1950s, the QCB managed a system of what it called "allocations to producers and consumers" by which it worked to match production in the major coal producing regions to demand in each region, also paying attention to the quality of coal needed by various consumers.[545] The Brisbane area relied largely on West Moreton coal, but this needed to be supplemented by coal freighted from the central region. The central region, with the Callide and Blair Athol open cuts the major producers, also supplied coal to the northern region of the state. The QCB said that it tried to ensure that the maximum quantity of coal was supplied from each region to its local consumers provided the local coal also met certain quality standards. Recognising the freight costs involved in moving coal between regions, and the demands on the railway system, the QCB said that it attempted to keep the inter-region shipments to a minimum.[546] The QCB also set prices for various grades of coal for each of the regions in the state, with for example a fixed price set for all coal produced in the Bundamba area and a slightly higher price for all collieries in the Rosewood area. There was a significant variation in prices throughout the state so that collieries could earn a reasonable rate of return. The Chillagoe mines in 1954 received the equivalent of just over $9 per ton for their large coal, while prices in other regions for underground collieries averaged between $5 and $6 per ton. The Callide open cut mine received only $2.20 per ton, and the Blair Athol mines from $1.70 to $2.80 per ton, reflecting the much lower cost structure of these mines.[547]

Power station plans give hope to the NSW industry

After the severe cutbacks by the Greta seam mines, the expansion plans of the Electricity Commission of NSW were a major source of hope for the State's coal producers and mine workers. The Commission was established in 1950, and it inherited the major power stations in the state, including in 1952 the Bunnerong and Pyrmont stations owned by the Sydney County Council. Other power station plants and transmission lines were transferred to the Commission in the early 1950s. In January 1953 the Commission gained ownership of the 4 coal fired power stations previously owned by the NSW Railways in Sydney (White Bay and Ultimo), Newcastle and Lithgow.

Within its first couple of years, the Commission was engaged in a number of new power station projects, including some diesel powered, to overcome the severe shortages of power that the state had been experiencing after the war. Its first major new power station was Wangi on the southern edge of Lake Macquarie. Wangi was developed as a 330MW station, and came online between 1956 and 1958. Next came Tallawarra, situated on Lake Illawarra south of Wollongong, a 320 MW station that was commenced in 1954 and that was in full operation by 1961. Wallerawang, near Lithgow, was the third major station developed in the 1950s, the small A station of 30 MW completed by 1959, with the B station of 120 MW completed in 1961, and the two 500 MW units of C station completed in 1976 and 1981.The growing power demand meant that these new stations were creating a much larger market for coal to fire their boilers.

The NSW Electricity Commission acquired it first coal mines in 1958 when it purchased the Newstan and Newcom collieries from the JCB. It would go on to take over existing mines from the NSW Government's State Mines Control Authority in the 1973 and become a significant miner in its own right until its mines were sold to Centennial Coal in 2002. The mines acquired in 1973 were Awaba, Wyee and Munmorah mines located south of Newcastle, and Liddell in the upper Hunter Valley. By 1958, the Commission's coal usage had grown to almost 3.5 million tonnes, and it was forecasting that figure to increase to 3.8 million tonnes by 1960, and then to jump to 6.5 million tonnes by 1971 as new power stations built on the coalfields came online. Much of its coal requirements came from its own captive mines or from mines operated under contract. The steel industry was also growing, and with it the demand for coking coal. From an estimated 4.7 million tons in 1960, demand for coking coal was forecast to grow to 6.4 million tons by 1970.[548] In Queensland demand for electricity was still growing slowly in the 1950s and it would not be until the 1960s that new power stations were commissioned. Callide A, Swanbank A and Collinsville stations would all commence generating between 1965 and 1968, providing a much needed boost to the local coal market.

Mine workers' health gets greater attention

While driving higher production was the major focus of the JCB in its early years, another of the JCB's priorities in the 1950s was the health of mine workers. As noted earlier, a JCB survey of around 1500 mine workers in 1948 had revealed that one in six had pneumoconiosis (black lung disease), and a similar number had recurrent bronchitis or chronic lung disease. Black lung and other diseases were clearly blights on the industry and needed to be eradicated. A major change was introduced by the JCB in 1950 when it mandated that all new employees coming into the industry were required to have a medical examination, and periodic examinations were also mandated for all workers in the industry. New entrants into the industry could now be screened for medical problems, and prevented from employment if they were not provided with a medical certificate. Ongoing health checks would now allow progress to be monitored. The JCB's medical centres in the major coal areas were now also providing medical treatment for workers affected by lung disease. By mid-1950, the JCB's medical centres had examined 4,000 mine workers and workers seeking employment in the industry; examinations were not just for miners, but also included those in clerical and administrative positions.[549] The JCB was also focussed on reducing the problems of dust through promoting good mining practices, and in particular ventilation of working areas at the coal face.

The JCB's medical program was led by Dr W E George, who had been chief medical officer of the Medical Bureau in Broken Hill for 21 years, and had won widespread support for his work from companies, unions and the community. At a community farewell in Broken Hill, George Fisher (later Sir George and who went on to lead Mount Isa

Mines), said that Dr George had gained the confidence of the Broken
Hill miners and the companies, and had examined perhaps 10,000 men
during his term in the city.[550] Dr George was appointed as chief medical
officer of the JCB in August 1947. The JCB also established the Standing
Committee on Dust Research in 1954, with the various sectors of the
NSW industry represented on that committee; this committee is still
operating and its function now is to review airborne dust concentrations
in the industry and the results of research into dust control.[551]

The NSW Department of Mines also began to focus more on dust
problems in the early 1950s, although its major concern was in relation
to the potential for coal dust to cause an explosion. Stone dusting,
using limestone to render coal dust inert, had been introduced into the
industry in the 1920s, particularly following the explosion at the Bellbird
colliery near Cessnock in 1923 and the report of the Royal Commission
on Safety in Mines in 1926. In 1926 the Coal Mines Regulation Act was
amended to require all roadways and all new workings to be treated so
as to remove the danger of coal dust; the Act also required areas to be
stone dusted which carried electric cables to remove any chance of a
spark igniting the coal dust.

While there was obviously improvement over the years, major problems
still existed in the 1950s. In its 1951 annual report the Department reported
the results of the Department's sampling of dust levels in collieries
in 1950 and again in 1951, with multiple locations sampled in many
collieries. In 1951 the Department visited 122 collieries and took dust
samples in 381 locations; only 62% of the locations sampled were found
to be complying, or almost complying, in terms of dust control. In 1952
the Department reported that the percentage of locations complying
or almost complying had risen slightly to 62%.[552] Some of the mines
visited were naturally damp and many locations therefore posed little
danger, however the Department found that, excluding those locations,
a small number of collieries accounted for a high share of the complying
locations. In 1951 for example, only 12 collieries accounted for 42% of
all locations which were fully complying. For two well-known collieries
mining the volatile Greta seam, one had no locations fully complying,

and the other had only two fully complying locations.[553]   Clearly a small number of collieries were on top of the problem and were controlling dust throughout their operations, but many others were falling well short of what was required. In 1951, the Department warned colliery owners that they needed "to make arrangements to ensure that plentiful supplies of stone dust are available." Stone dust had been difficult to obtain prior to 1952, but with supplies now readily available, the Department was now putting the industry on notice.

The focus on the control of dust and treatment of the workers with lung problems would prove to be a major success over the coming years. By the early to mid-1960s, the JCB's regular medical examinations showed major improvements in the health of mine workers. In the years from 1963 to 1965 the JCB examined just over 6700 mine workers; pneumoconiosis was detected in 2.9% of those workers, and recurrent bronchitis in 5%. These figures were of course still unacceptable, but demonstrated the progress which had been made since the initial survey in 1948.[554] The 1963-1965 examinations showed a slight reduction in cardiovascular and renal disorders compared with 1948, perhaps indicative of a range of causal factors including those not related to working conditions.

Minesite amenities improve

In the early post war years, the standard of facilities at mine sites was generally poor to abysmal. In its first annual report, the JCB accused the producers of failing to recognise the need to upgrade amenities and facilities underground and on the surface: "In the field of so-called pit amenities, the Board has had to contend with an attitude of mind on the part of most colliery proprietors which is entirely unreal. With their perspective clouded by the traditional low standards of the industry, many colliery proprietors have been unable to regard expenditure designed to improve working conditions and pit top efficiency as anything but a waste of money."[555] The JCB surveyed mines in 1947 and later documented the number of mines where certain facilities were up to standard.

In 1947 there were 155 coal mines in NSW. Only 3 mines had bath and change houses which were up to standard; sanitation facilities on the surface were up to standard in only 14 mines; and the surface water supply for bathing was up to standard in only 63 mines. Underground the picture was even worse: water reticulation was up to standard in only 7 mines, there were no mines with crib rooms up to standard, and sanitation in only 2 mines was up to standard. However there was a major improvement over the period to 1952 by which time the number of mines had risen to 176. The corresponding numbers of mines with facilities up to standard was: bath and change houses 106; surface sanitation 81; surface water supply for bathing 151; underground water reticulation 84; underground crib rooms 53 and underground sanitation 24. In 1952 a number of mines also had facilities under construction, including 41 bath and change houses and 22 surface sanitation facilities.[556]

Information on the facilities at Queensland mines for this period is lacking in the QCB annual reports. The Powell Duffryn report in 1949, in noting a relatively low rate of absenteeism in the Queensland coal industry (around 10% in undergrounds and as low as 3% in open cuts), said that "the lack of many so-called amenities, many of which in other industries would be regarded as necessities, has not so far been allowed to be used as an excuse for irresponsible stoppages, though it is almost certain that it has been responsible for the difficulty in recruiting labour in many of the less favourable areas."[557] The report also noted that in 1928 the provision of baths on the surface of mines had been mandated for all mines which normally employed 5 or more workers underground (although an exemption was possible if the provision of these facilities was impractical); and the statutory standards for bath-houses were lifted in 1947, with Powell Duffryn saying that "improved bath-houses have been and are being erected at many mines" but that the problem of supplying water for bath-houses, while a difficult one, needed to be tackled.[558]

Bowen Basin coal pioneer Sir Les Thiess 1972. NAA A6135 K1/5/72/43

New dragline at Moura mine 1967. NAA A1200 L64026

9

The Queensland Giant Comes Alive

Sir Leslie Thiess - Queensland coal pioneer

The existence of extensive coal resources in Queensland had been recognised for many years, although the full extent was only starting to be understood in the early 1960s. The pioneers of the development of the post war coal mining industry in Queensland were Leslie (later Sir Leslie) Thiess, US company Utah Development International, and the Japanese company Mitsui. Leslie (or Les as he was commonly known) Thiess was the managing director of Thiess Brothers, a company which had its origins in a road building operation in the 1920s and 1930s and which was run by Les and his brothers. During World War Two Thiess Bros worked with the US Army and the Allied Works Council on roads, military camps and emergency airstrips around Australia. In 1944 the company made its first foray into coal mining when it won tenders to remove the overburden at one of the Blair Athol coal mines in Queensland and the Muswellbrook coal mine in NSW. Among many projects in the 1950s, the company won a contract to build two tunnels for the Snowy Mountains scheme, the first major contract on the scheme won by an Australian company.

The company lost the Muswellbrook contract in 1957 and then the contract with Blair Athol Coal and Timber Co. Thiess Bros also had a coal mine at Callide at the time supplying local markets, although the early Thiess operations at Callide were apparently not of a high standard. Tom Hiley, the Queensland Treasurer in the Nicklin Government, is quoted as saying that in relation to Callide, the Government needed a company like Thiess to be involved in the development of that mine. However, "in the early phases, Thiess's work at Callide was something of an engineering shambles. They'll tell you so themselves. They had holes everywhere and ore buried all over the place - and a succession of engineers in the process. They learnt everything the hard way, but they learnt it quickly and Les was straightforward and reliable to deal with. We knew the thing would go ahead and they'd end up on a sound engineering basis. They were above all practical people."[559]

However Les Thiess wanted to develop his coal operations for the export market and was "determined to look for (Asian) markets and managed to get several orders, one from Japan, another from South Vietnam. But there was no follow up in either case as it was clear that for export purposes, coal of a different quality was needed to mix with the Callide product. He had an instinctive feeling for the future of the industry and decided to finance personally a search in Central Queensland for this quality coal." [560] Thiess then employed Dr. Freddy Whitehouse. Whitehouse was "the best known consulting geologist in the state … If anyone could find the coal Les was after, it was this man…He set off now for the Bowen Basin in the vicinity of the town of Blackwater, one of the most likely areas documented by the British experts, Powell Duffryn, at the instigation of the Queensland Government in 1949."[561]

Les Thiess was initially interested in finding coal to blend with his Callide coal. Despite a significant drilling program, the area near Blackwater did not prove promising, but in 1957 Whitehouse discovered soft coking coal deposits at Kianga near the end of the south eastern branch of the Bowen Basin, followed in 1958 by discovery of medium quality coking coal at nearby Moura. These discoveries heralded the commencement of the massive development of the coking coal resources in the region, a region which would prove to be one of the most valuable and productive regions in Australia's history. However, if not for the involvement of Mitsui, the history of the industry in Queensland may have been different. Mitsui's initial contact with the Australian coal industry was in 1954, when NSW producers were interested in exporting coal to Japan. Mr I Kitahara from Mitsui visited to investigate the industry and arranged for samples to be sent back to Japan for testing. The coal testing proved unsuitable and apart from some soft coking coal, no major Australian coal exports were recorded for the next few years. However, as we saw earlier, the visit of the Japanese coal group who inspected NSW coal mines in 1958 soon led to firm contracts for the supply of coking coal from several NSW south coast mines. The group also visited Kianga where Thiess was developing a box cut (a small excavation), but this development had not progressed far enough for the coal seams to be revealed. The Japanese, not being able to actually see the coal seams, lost interest in Kianga. However on his return to Japan, Koichiro (Ken) Ejiri, the manager of Mitsui's coal division, who was a member of the mission, recommended to Mitsui and Thiess that there should be a follow up visit by Mitsui geologists. This recommendation was accepted, and a team of Mitsui geologists arrived in February 1959. The Mitsui team however found that the Kianga soft coking coal was of a type which at the time was readily available in Japan. What Japan needed was hard coking coal, very little of which it produced domestically, and which was largely sourced from the USA. The team however believed that hard coking coal potentially lay further north and recommended that a new drilling program be undertaken. The drilling was supervised by Dr. Whitehouse and was soon successful in finding hard coking coal.[562] Ken Ejiri would go on to become the chairman and managing director of Mitsui Australia from 1971 to 1974, president of Mitsui & Co from 1985 to 1990, and Mitsui chairman from 1990 to 1996.

Earlier in 1958 Thiess had been granted approval to commence open cut mining at Kianga, with the State Government stipulating that all coal produced was to be sold to markets outside Queensland. The first trial shipment to Japan was made in October 1959, and further trial shipments followed. These shipments were sent to Japan for testing by the steel mills, an important step to provide the Japanese with an understanding of the technical qualities of the coal, and its suitability for use in Japanese blast furnaces. In November 1961 Thiess was successful in obtaining a firm order from the Japanese steel mills for 2.4 million tons, this order marking the real beginning of the massive expansion that was to come in the Bowen Basin. In December 1959 the Queensland Government granted Thiess rights to explore and mine an area of 350 square miles (around 900 square kilometres) from Baralaba in the north to Theodore in the south of the coalfield. This was a massive area, and Thiess Bros had limited financial resources and mining expertise to undertake major new developments on its own.

Mitsui's involvement in the development of the Kianga and Moura areas from the early days also included providing Les Thiess and his staff with technical assistance. Mitsui advised Thiess that if the company was to develop the project on a large enough scale, they would need to find an experienced American company to partner them in the project.[563] Thiess did find such a partner in Peabody, a major US coal producer, and Thiess Peabody Coal Company Pty Ltd was formed in January 1962. Towards the end of 1960, Les Thiess returned from another trip to Japan with news that he expected his company to commence exports to Japan in 1961, with export volumes rising to reach 2.5 to 3 million tons a year by 1970. Thiess said that the JSM would need about 22 million tons of coking coal per year if they were to achieve their planned output of around 33.5 million tons per year of steel, with the potential for at least 5 - 6 million tons to be provided by Australia.[564] Thiess had to wait a little longer than expected, but in January 1962, the Queensland Mines Minister announced a new Thiess Peabody contract with the JSM for the supply of 3.4 million tons of coal over the next 7 years which included the 2.4 million tons agreed only a couple of months earlier. This was the largest contract yet signed with the Japanese and Thiess Peabody Coal was now the other major new player in the industry, along with Canadian company Placer which was focusing on the Burragorang Valley in NSW. In April 1963 Thiess Peabody installed a massive Marion walking dragline at the Moura mine, the largest of its kind at that time in the southern hemisphere, capable of excavating almost 1200 cubic metres of earth and rock each hour. A second dragline was installed in 1967.[565] In January 1963, Mitsui cemented its involvement in the Moura project, taking a 20% equity, and the company was renamed Thiess Peabody Mitsui Coal, or TPM. TPM went on to win further contracts with the JSM, with a new contract signed in 1964 for the supply of 29.5 million tons over 13 years. That contract was worth $240 million, a massive sum in the early 1960s, but which would be dwarfed when Utah and the JSM signed export contracts in the late 1960s.

Thiess Peabody and the Queensland Government formalised the

approval for the project through the 1962 Thiess Peabody Agreement Act. Apart from provisions for special mining leases, one of the key provisions of the Act was the requirement for the company to build its own railway line from Moura to the coast at Gladstone, a provision which would change in 1965. The Government at the time was happy for the company to take on that task, as Mines Minister Ernie Evans said: "We have had enough to do with building railway lines. We are losing too much on existing lines. We believe it is much better to leave that to private enterprise..." [566] But that private line was not built; instead, in 1965 the Thiess Peabody Agreement Act saw key provisions of the 1962 Act re-negotiated, with the Government now taking responsibility for the construction and operation of the railway line from Moura to the port. The Government, and in particular Gordon Chalk as Transport Minister (and later Treasurer), had now started to understand the potential for the Queensland Railways to make real money from the new coal developments. The new Act also provided for increased, although still very modest, royalties to be payable to the Government, but its most innovative provision was the introduction of a security deposit payable by the company to the Government. That security deposit was then used to pay for about half of the cost of the railway and new rolling stock, relieving the financial strain on the State Government. Later legislation would see security deposits effectively pay for the whole of the cost of new rail infrastructure. With rail freight rates for TPM's coal set to not only cover the operating costs of the line, but a handsome return on capital for the Government, the Government was now positioned to benefit handsomely from the expanding coal industry.

The expansion of the Moura mine and its success in winning major contracts with the JSM also caught the attention of the Joint Coal Board in NSW, with the JCB anxious that these contracts would undermine the modest success which NSW exporters had achieved in the late 1950s and early 1960s. In an appeal to the industry and governments, the JCB said in 1962 that it was "In the best interests of the marketing of export coal from the two states (that there should be) the closest possible association between Queensland and New South Wales ...so as to ensure that misunderstandings do not arise leading to unnecessary competitive price reductions to the disadvantage of all concerned." The JCB went on to argue that the only winners from price competition and the "ultimate profit instability which might arise (and which) ... could have unfortunate consequences ... (would be) those in other countries supplying the export market." The JCB said that it had no doubt that our principal market - the Japanese steel industry - did wish to see such an outcome. "The Australian interests concerned in New South Wales and Queensland will need to protect themselves by fostering their mutual interests and avoiding conflicting policies."[567]

The JCB of course was keenly aware that its counterpart in Queensland, the Queensland Coal Board, did not control the development of the export sector of the industry or have any power over export contracts and prices. Together with the NSW producers, the JCB understood the threat which was emerging from north of the border. The JCB's reference to the Japanese steel industry not wishing to see price competition and the resulting "profit instability" was perhaps well-meaning, but would prove to be misguided. The JSM would understand quite clearly the competitive advantage offered by the new Queensland Bowen Basin mines, and would ruthlessly exploit it in the coming years.

Utah explores the Bowen Basin

The Utah Construction Company was founded in 1900 and in due course became a major construction company in the USA and internationally, with activities including the construction of railways and dams, one of which was the massive Hoover Dam on the Colorado River in the USA. After World War Two, Utah moved into coal, iron ore and uranium mining. By the 1960s its name was Utah Development International, and its interests in Australia included construction of the Eildon Dam in Victoria, Snowy Mountains Scheme projects, and the Lake Moondarra Dam in Queensland. In the 1950s Utah and Cyprus Mining Corporation combined to develop an iron ore mine in Peru, and they also purchased a fleet of bulk ore carriers to ship the iron ore to their customers in Japan.

In 1959, Utah recruited Richard Ellett, an American geologist who had spent time in the 1950s investigating mineral prospects in Australia. Ellett moved to Melbourne in October 1960 to establish Utah's Australian exploration operations, his brief being to focus on exploration for minerals which would be suitable for large scale mining, with the emphasis on iron ore and coking coal.[568] Being already involved in the iron ire trade, Utah was a company which was ahead of most others in understanding the potential for growth in the Japanese market for iron ore and coal. The company also believed that the Australian Government's ban on the export of iron ore would soon be lifted, and that with a dramatic increase in Japan's steel industry expected in the 1960s and 1970s, the potential for iron ore and coal supply was massive.[569] An added factor in favour of developing new non-US coal projects was that Japan was also looking to diversify its coal suppliers to reduce its heavy reliance on the east coast US coal producers. Utah's iron ore work led to a feasibility study for the development of the Mt Goldsworthy deposit in the Pilbara region of Western Australia, and in due course its construction by a joint venture involving Utah, Cyprus Mining and Consolidated Goldfields of Australia. Mt Goldsworthy was the first of the huge iron ore projects to be developed in the Pilbara.

Ellett and his boss, Weston Bourret, were not optimistic about the potential for large scale open cut coal development in NSW, and soon looked to Queensland. But they had a major problem – they did not have a detailed knowledge of coal geology or experience with coal. But Ellett did not let the lack of knowledge deter him, and, being aware of the Bowen Basin, applied for a prospecting authority on land near Thiess's leases at Kianga. This was no doubt a smart move, although Ellet said that: "It did not take a great deal of shrewdness to realize that the best place to find coal was within the same stratigraphic measures adjacent to a known coal deposit."[570] Bourret made three trips to Japan to find out about the technical specifications for the coal the Japanese steel mills were seeking. He did this with the assistance of Mitsubishi, the other giant Japanese trading and industrial company, which, like its rival Mitsui, also had a long history as a coal miner in Japan. Bourret

also met with staff of the giant Japanese steel company Yawata Steel
(which later merged with Fuji Iron and Steel to become Nippon Steel
Corporation).[571]

The potential of the Bowen Basin had been recognised for many
years, but only in general terms. Ludwig Leichhardt, the famous explorer,
reported finding coal on the banks of the Mackenzie River near Blackwater
in 1845. Attempts were made to mine coal in the area from the 1890s,
but with limited ongoing success. As was noted earlier, after World
War Two the Queensland Government commissioned UK firm Powell
Duffryn Technical Services to review the potential for development of
the coal industry in Queensland. The company referred to many studies
and reports, noting for example that at the northern end of the Basin,
there may have been one billion tonnes of "workable deposits" of
coal within an area of 120 square miles.[572] The Powell Duffryn report
concluded that even larger deposits were likely to exist elsewhere in the
Basin, and recommended a detailed drilling program to obtain more
accurate information. However little happened in Queensland in terms
of new development outside the established coalfields until the historic
discoveries by Thiess Bros in 1957 and 1958.

In 1961, Ellett expanded his team with some key staff, including Don
King, a geologist recruited from Rio Tinto. King and his colleagues would
prove to be very valuable acquisitions by Utah. King and his team first
carried out drilling in July 1961 in an area near Theodore and intersected
good quality steaming coal. Investigations moved on through Baralaba
and Bluff to Blackwater, where the company believed the potential
justified a detailed drilling program. This was successful and it was at
Blackwater that the team found good quality coking coal. On the day
following ANZAC Day in 1962, Utah's geologists intersected a coal seam
around 7 metres thick at a depth of only 33 metres. Further drilling in
the north western area of the Bowen Basin soon found the rich seams at
Goonyella. Utah was now in a position to move ahead, and the company
was awarded a "Proclaimed Area" of 6200 square kilometres, a huge part
of the overall Basin. It also should be noted that the Commonwealth and
Queensland Governments played an important part in the exploration of

the Bowen Basin through the work of the Bureau of Mineral Resources and the Geological Survey of Queensland. These organisations, in cooperation with Utah, carried out extensive geological mapping of the Basin which assisted to improve the understanding of the region's geology.[573] Don King later resigned from Utah and was recruited by Mines Administration, a company which was part of petroleum group AAR Ltd, where he made further major discoveries.

Utah was now in the box seat to become the leading coking coal exporter, not only in Australia, but in the world. But Utah's experience in Australia would be controversial, with the company accepting prices from the JSM well below what many Australians saw as reasonable, but with the company's mines still proving to be very profitable.

Utah begins to mine

By late 1962, with the exploration program proceeding and proving extremely positive, Utah's head office in San Francisco gave the go-ahead to expand the program which would lead in due course to development of the Blackwater, Goonyella, Norwich Park and Saraji mines. In December 1964 the Queensland Government granted Utah exploration rights (through an Authority to Prospect) covering a massive 2500 square miles (almost 6500 square kilometres), with the Authority incorporating some earlier rights granted to the company. Utah now had the sole rights to what would prove to be the best coking coal reserves in the world for a period of 5 years from 1 June 1964. The company however would be required to give back a third of the area every two years, but its Authority gave it exclusive rights to secure mining leases, consolidating "Utah's grip on the lion's share of the most promising areas of the Bowen Basin, giving it a segmented strip of country stretching 150 miles from Blackwater to Goonyella."[574]

Why such a huge area was granted to one company has been a controversial issue. Mines Minster Ron Camm (who succeeded Ernie Evans, the Minister responsible for granting the Authority) later said that the area was granted to Utah "…to encourage big development.

It was very hard in those days to get anyone to go out and spend the amount of money that Utah was prepared to spend. But they weren't prepared to spend those amounts of money unless they were sure of getting something. Within these areas there were tremendous deposits of carbon, which the Department knew were there but didn't know in what quantity. Utah, give them their due, were the ones who proved the quantity of coal that was in that area."[575]

Blackwater was the first of the Utah mines developed, and was first for a number of reasons. Apart from being the first area which was subject to detailed drilling, Blackwater was also a mine which was able to be developed for the very modest cost of only $30 million.[576] It did not need a huge investment in railways or ports, and was able to use the existing rail system and export through the port of Gladstone.[577] The basis for the Blackwater project was a contract between Utah and the JSM signed in November 1965 for the supply of 13.5 million tonnes over ten years, commencing in 1968. The sales agreement with the Japanese was amended in 1968 to just over 21 million tons, and again in 1971 to just over 28 million tons.[578] With Mitsui an integral part of the Thiess operations, its arch rival Mitsubishi was now a key player in the Utah developments from the mid-1960s. The various companies in the Mitsui and Mitsubishi groups, the two largest of the Japanese "sogo shosha" (trading companies), would continue to play a major role in the development of the Australian coal industry, one which still exists today. Mitsubishi, with its vast operations including commodity trading and its understanding of the Japanese steel and other industries would provide a vital link for Utah to the emerging Japanese coal market.

The development of Blackwater also involved an upgrading of the railway line to Gladstone and the purchase of 16 diesel locomotives and 350 aluminium wagons by the Queensland Railways. The details of the Blackwater railways agreement were not made public, but, apart from the Central Queensland Coal Associates mining developments under the 1968 Act, it was the model for subsequent agreements between the Government and coal companies.[579] It is safe to assume however, that the Queensland Government would not have suffered financially from the arrangements.

Blackwater was formally opened in May 1968 by the Queensland Premier, Jack Pizzey, and Utah International's chairman, Mariner Eccles. By 1968-69 the mine had lifted its production to around 1.3 million tonnes, rising to around 2.7 million tonnes in 1969-70. The contracts underpinning the development of Blackwater were the largest ever entered into by the Japanese steel mills, but by 1969 would look quite modest in comparison with contracts which Utah and the JSM struck for the company's Goonyella and Peak Downs mines.

The agreement between the Government and Utah provided for steaming coal from the mine, which would otherwise have been discarded along with overburden, to be mined and stockpiled for later use in power generation. The steaming coal at that time had little commercial value and Mines Minister Ron Camm, aware of the steaming coal wasted at Moura, wanted to avoid a repetition at Blackwater. The Government agreed to pay Utah for the costs of stockpiling the coal, and was rewarded during the 1970s when the big new Gladstone power station came on stream and was able to draw on the Blackwater stockpiled coal.[580]

The development of the Bowen Basin would change the region fundamentally over the next twenty years, with existing towns such as Blackwater and Emerald to grow rapidly, and new towns such as Moranbah to also emerge. Blackwater in the 1960s was a very small country town, servicing the local farming and grazing community, and was a stop on the railway line which ran from Rockhampton west through Emerald and Longreach. The whole Duaringa Shire which included the town of Blackwater boasted a population of barely 2000 in 1966. By 1976 the Shire's population had grown to around 7700 and Blackwater's to around 4600. Blackwater's infrastructure in the mid-1960s was also fairly primitive, as evidenced by an anecdote later told by Utah's Dick Ellett. Ellett said that when a senior executive from Utah's San Francisco visited Blackwater, he was invited to take the "executive suite" at the local pub. "Unfortunately, he did not realize that the only window in the room opened on to a small generator that produced power for the pub. With the puffing and chugging of the generator, and through inhaling its fumes, our visitor, as we had anticipated, obtained very little sleep and had a headache the next morning. It was not that we did not like him, but as a joke thought it would be helpful if those with 'cushie' jobs in San Francisco experienced some of the discomforts of the field men!"[581]

In 1968 the Queensland Government passed legislation to give Utah and Mitsubishi the power to proceed to develop new mines at Goonyella, Peak Downs, Saraji and Norwich Park. Utah and Mitsubishi established a new company, Central Queensland Coal Associates Limited (CQCA), to be the vehicle to develop the suite of massive new mines. The Central Queensland Coal Associates Agreement Act 1968 and the associated agreement provided for a 5 year authority to prospect covering an area of almost 3500 square kilometres. The area reduced each year to reach around 780 square kilometres and during this period the company was able to determine which areas it wished to target and mine. The agreement allowed CQCA to apply for up to 4 huge leases, with the total area not to exceed 175 square miles (around 450 square kilometres), and the volume of coking coal allowed to be mined was limited to 150 million tons, with a further 150 million tons provided that this additional tonnage did not exceed 30% of recoverable reserves.[582] The area of the leases was huge, although the restrictions did mean that Utah had to forego certain attractive areas such as German Creek which would later be exploited by other companies.

Utah's massive Goonyella and Peak Downs contracts

On his return from a visit to the USA and Japan in 1968 the Queensland Treasurer, Gordon Chalk, announced that Utah, Mitsubishi and the Queensland Government had agreed in principle to develop a major new mine at Goonyella, around 200 kilometres south west of Mackay. The new project would involve the construction of a new rail line to the coast, and a new township for a population of 500, with the coal to be exported through Mackay or a new port at Hay Point. The commencement of the project was subject to firm financial agreements and export contracts. The project was expected to export 4 million tons of coal per year.[583] The Queensland Government, having learnt from its experience with the TPM Moura project and Utah's Blackwater project, said that Utah would be required to provide funding for the cost of the rail line through a security deposit which would be drawn down to pay for construction. If Utah performed as required under the agreement, the security deposit would be refunded over a 12 year period.

The royalty arrangements for the project now seem hard to believe. The initial royalty rate was 5 cents per ton, the standard rate which applied at that time. However it would prove to be difficult to justify and the Government took a unilateral decision to increase the rate in 1974. The Government however was not too concerned with the royalty rate in 1968 as it was looking to build a de facto tax or royalty into the freight rate which the company would be required to pay for the use of the rail line. Chalk said that as well as escalation clauses being included in the agreement to provide for rises in the costs of operating the rail line, "an attractive profit margin" would be built into the freight rate.[584] As it would become clear in later years, the Queensland Government was to become expert at extracting significant taxation revenues from the major coal projects which were to be developed in the 1970s and 1980s. The royalties paid by mining companies may have appeared modest, but when the profits earned by the Queensland railways were included, the State became a major beneficiary from the boom in the export coal industry.

In January 1969 came the formal announcement from Utah and Mitsubishi that they had signed a massive contract with a group of 14 Japanese steel mills, gas companies and chemical companies, to supply 85 million tons of coal over a 13 year period, worth around US$1 billion. Of the 85 million tons, just over 50 million was to be supplied from the new Goonyella mine commencing in July 1971, and the remainder from a new Peak Downs mine to be supplied from 1972. Utah owned 85% of the new CQCA joint venture vehicle, and Mitsubishi 15%.[585] The new company's developments did not include the Blackwater mine; that remained 100% owned by Utah, although the ownership was later restructured to provide a new locally listed company, Utah Mining Australia Limited, with 10% of the equity. The investment by Utah and Mitsubishi in the new Goonyella mine and the new township and the new port at Hay Point was expected to total around US$112 million, with the port capable of handling ships of up to 100,000 tons.

The price for the coal was to be US$11.98 per ton, fob Australia, a price which appeared to compare unfavourably with the US$14 per ton which the USA exporters were expected to receive once the Utah Mitsubishi contract became effective. Not only was there this US$2 difference in the fob price, but the shipping freight cost of the US coal was around US$4 per ton, compared with an expected US$2 per ton or less from Australia. However, in announcing the contacts Mitsubishi, which had assumed a major role in the negotiations, said that it had made "strenuous efforts" to keep the price below US$12 a ton, as this price would become the standard price for coal supplied from other sources to Japan.[586] Mitsubishi appears to have been referring to potential new contracts with Canadian exporters whose cost structure was higher than Utah's and for whom the export price would be difficult to match.

The 85 million ton contract was believed to be the largest export coal contract ever signed and represented a new era for the Bowen Basin and Queensland coal industry; it dwarfed the contracts which TPM had won, and generated a good deal of angst from NSW coking coal producers. The issue of whether the prices were fair and reasonable would become a major issue for Utah, the JSM and the Australian Government in the 1970s.

Other Bowen Basin developments also taking shape

The Utah developments in the late 1960s were the most significant mining developments in Australia. However the Bowen Basin also saw other developments and plans, although some of these would take many years to come to fruition. Thiess Bros, which was unsuccessful in its earlier exploration near Blackwater, had found quality hard coking coal just south of Utah's Blackwater mine, and was planning a 5 million tonnes per year export operation. With its Moura mines having expanded to produce almost 3 million tonnes per year for export, Thiess then moved to develop a new mine at South Blackwater. In April 1969, a contract was signed with Yawata Iron & Steel, acting on behalf of a number of mills, for supply of 15 million tonnes over 15 years, with deliveries to begin in 1970. The mine was developed to export through the port of Gladstone, and Thiess also proceeded to build a 40 kilometre rail line to connect the mine to the main central Queensland system.

Following the pioneering work of Thiess Bros and Utah, a number of other companies started to explore in the Bowen Basin. BHP found good quality coking coal at Yarrawonga, south east of Blackwater, and by 1969 was discussing with the JSM the potential for export of 2-3 million tonnes per year, with a similar quantity going to its own steel mills in Australia. Clutha Development commenced exploration in 1966 at Sirius Creek and then commenced a trial mining phase. By the late 1960s, Clutha was planning to develop three underground mines with a total production of 5 million tonnes per year, and which would be connected by a new rail line to an export terminal in the Shoalwater Bay area capable of accommodating ships of 200,000 deadweight tonnes. Clutha's Sirius Creek proposal however was rejected by the Queensland Government as the company was planning a rail line with a capacity of 20 million tonnes per year. Its own mines would produce only 5 million tonnes and the line would therefore carry coal for other exporters. Clutha wanted to enjoy some of the profits from rail freight, however the Queensland Government was not keen on giving up the potential to earn significant revenue through its control of the rail system and the setting of freight rates. Mines Administration (part of the Associated Oil Group) had found coking coal at Bluff, a resource suitable for underground mining, but which at that time was not viable, with open cut mines being much cheaper to operate.

By 1970 the major mines producing in the Bowen Basin were the TPM Kianga/ Moura mines (almost 3.5 million tons for the year), Utah's Blackwater open cut and underground (3.1 million tons), Thiess' South Blackwater (just over 0.5 million tons), and Utah's Goonyella (which began production in the 1970 calendar year, producing 0.2 million tons in the January to June period). BHP's Leichhardt mine was in a trial mining phase and produced 24,000 tons.

Pressure on NSW to nationalise the industry

While the massive development of the Bowen Basin was beginning in the early 1960s, the mine closures and retrenchments of the late 1950s and early 1960s saw pressure on the NSW Government to do something to "save the industry". The Miners' Federation had not succeeded in getting rid of the JCB or making any major changes to its policies, so its efforts turned to convincing the Labor Party of the need to get the NSW Government to act. The trends in the domestic markets for coal were pretty clear. Between 1954-55 and 1959-60 the consumption of coal nationally by the railways had fallen by 36%, and consumption by the producers of town gas was down by almost 19%. These reductions were balanced by the 20% growth in coal use by the electricity industry and by the 25% growth in use by the steel industry.[587] These market trends, while positive for some producers, meant death or poverty for others, particularly the Greta seam producers around Cessnock and some of the producers in the NSW western district which had been major suppliers to the railways.

In 1963, after consultations between the members of the Parliamentary Labor Party, the NSW State ALP Executive and the NSW Labor Council, the Acting Premier, Jack Renshaw, announced to the annual NSW Labor Conference in June that the Government would appoint a Parliamentary Committee to consider the merits of government ownership of the industry.[588] A few days later, in formally announcing the decision, Renshaw seemed to have broadened the scope of the reference to the Committee, which would be appointed after the Parliament had resumed in August. Renshaw said that the Committee would be asked to investigate all aspects of the coal industry including whether the Government should take control of the coal industry. The coal producers reacted with concern, Sir Edward Warren saying that "I am sure that Mr Renshaw could not be serious in putting forward this piece of doctrinaire nonsense." Warren said that the industry was already so efficient that it was competing with American coal in overseas markets. He stressed that further investment was needed to make the industry even more efficient and that talk of nationalising the industry would frighten potential investors away.[589]

Later that year, the issue became embroiled in the Federal election, when the Minister for National Development, Senator Spooner, gave a speech in Sydney in which he said that the NSW Government was planning to investigate the feasibility of nationalising the industry. The NSW Premier, Bob Heffron, who was overseas in June when Renshaw made the initial announcement, slammed Spooner for being "extremely irresponsible" for suggesting that his government was determined to nationalise the industry.[590] Spooner had referred to Heffron's letter to Prime Minister Menzies in August which contained the terms of reference for the Parliamentary Committee, the first of which was to investigate the feasibility of nationalising the coalmining industry in NSW, and if feasible, the terms and conditions on which that industry should be nationalised. The Committee's second term of reference was to investigate the possibility, short of nationalisation, of instituting an effective system of control over the production, marketing and distribution of coal. Also included in the terms of reference for the committee were consideration of ways to increase the domestic consumption of coal, and the possibility of large scale production of by-products from coal.[591]

The terms of reference for the Joint Committee of the Legislative Assembly were approved in December 1963 and the work of the Committee finally got underway in 1964. Chaired by Mines Minster J B Simpson, the Committee held a number of hearings at which it received evidence from key players including the JCB chairman, and it published a progress report later that year which contained the evidence presented, including the details of the hearings. While the feasibility of nationalising the industry was the first of the terms of reference, it is doubtful that the Committee had any real intention to pursue that option as the way forward for the industry.

As part of his evidence, JCB chairman Bernard Hartnell referred to the fact that the export sector of the coal industry in Queensland was not under the effective control of the Queensland Coal Board. The difference in the regulation of the industry in the two states would become a major issue in the 1970s and 1980s, but for the early 1960s, with the Queensland export industry just beginning to expand it was not yet a critical issue. In Queensland the export sector, led by Thiess and Utah, was in its infancy, but the focus of the QCB was on the mines which supplied the domestic market. It was the Queensland Government, led by its Premiers and Treasurers, which would encourage the growth of the export sector, including through various Acts of Parliament.

In May 1965 the NSW Labor Government lost the election and the Coalition led by the new Premier Sir Eric Willis gained power. The work of the Joint Committee lapsed and the new Government was not interested in reviving it. Nationalisation was certainly off the agenda, and it would not be until Labor won power at the Federal level in 1972, that major new controls would be imposed on the industry, this time in both NSW and Queensland.

NSW ports prove a major bottleneck

In 1960 the prospects for the NSW south coast producers were looking up, with the potential for significant exports to Japan. Export production was also assisting to limit the extent of job losses in the industry caused by factors such as mechanisation, with an estimated 850 jobs attributable to exports in 1959-60.[592] The JSM were expected to commence negotiations around mid-1960 with Australian exporters for a major ten year contract for coking coal, and National Development Minister Spooner left for a visit to Japan in June for discussions with the Japanese. Japanese industry leaders were reported to have said that the JSM hoped to import around 16 million tonnes per year of coking coal per year, with a significant proportion to come from Australia if transportation and port facilities could be modernised.[593] The NSW Government was now under pressure to invest in upgrading port and rail capacity, with all the coal terminals and the rail system owned and operated by the Government.

The capacity of NSW ports was a problem which had been well known in the industry and in government circles for some years. In 1959, the senior Japanese consul in Sydney addressed a meeting in

Wollongong and pulled no punches, saying that the inadequate facilities at Port Kembla could jeopardise the Australia Japan coal trade.[594] Shortly after the meeting, the chairman of the Southern Colliery Proprietors' Association, G W Perdriau, noted that the coal companies in the southern district "had spent many millions of pounds in the last five years on mine modernisation" and that investment by the companies was continuing in response to growing local and overseas demands for hard coking coal. However he criticised the Railways and the Maritime Services Board for being uncooperative on the question of the port's loading facilities, adding that "Many approaches have been made to the official bodies concerned to secure an improvement in coal shipping facilities in Port Kembla".

While US coal exporters were employing ships of up to 35,000 to 40,000 tonnes to carry their coking coal to Japan, the major ports in NSW could only accommodate ships of 10,000 to 12,000 tonnes. It was clear that Port Kembla, Newcastle and Balmain would all need to be upgraded, with a new jetty in Port Kembla, deepening of the harbour in Newcastle and better facilities at the Balmain terminal all required. Investment in the rail links to the ports and rolling stock was also needed. In May 1960 the NSW Premier announced that his government would spend "several million pounds" on upgrading Port Kembla and Newcastle, with a new pier at Port Kembla and a deepening of the entrance to Newcastle to approximately 11 metres to cater for larger capacity vessels. This announcement followed closely on criticism from Federal National Development Minister Spooner that NSW was not doing enough to improve its port facilities.[595]

The entrance to the harbour in Newcastle was less than 8 metres deep and thousands of tonnes of rock would have to be blasted away to achieve the desired depth. Despite the Government's announcement, it would not be until January 1962 that a contract was let for the work. In Port Kembla, work on constructing a new loader was begun in 1962, with the loader commissioned in 1964 with a modest capacity of 2 million tonnes per year. The Port Kembla loader had been funded by the NSW Government, with the Federal Government also providing financial assistance. Work was also carried out on new wharves. At Balmain in Sydney harbour, a new loader was completed in 1963, although the Government caused coal exporters to erupt when it announced an almost 50% increase in loading charges. Following protests from the industry, the increase was subsequently trimmed.

The upgrading in port capacity had been sorely needed. In November 1963 Robert Askin, then NSW Opposition Leader, attacked the NSW Government, noting that while the Federal Government had provided financial assistance to help the upgrading of the port facilities, the NSW Government had failed to work quickly enough. Askin said that Japanese industrialists had recently told him that the delay in NSW in NSW to enable large coal ships to access its ports had forced them to turn to other suppliers.[596]

While the entrance to the port of Newcastle had been deepened, the port's other major constraint was the lack of coal loader capacity. In 1967 a new coal loader in Newcastle (the Basin loader) commenced operating with a capacity to load 7 million tonnes per year for the export market. That was a major improvement, but would soon prove far from adequate. A private company, Canwan Coal, developed additional stockpile capacity on a site now occupied by the Carrington operations of PWCS, and this raised the export capacity of the port to 11 million tonnes. The NSW Government also carried out deepening of the channels in the harbour during the 1960s, work which was essential to cater for the larger ships which could enter the harbour.

By 1969, the outlook for exports to Japan was stronger than earlier in the decade. NSW Mines Minister Wal Fife visited Japan that year and was told by the JSM that they wanted to double their imports of NSW coal by 1975 to around 24 million tonnes, compared with the 12 million tonnes which they had bought in JFY 1968. In fact the JSM were reported to be looking to the NSW Government to invest in infrastructure to not only cater for the approximate doubling in exports to Japan, but for an even greater increase as they said that the tonnage could be higher than the 24 million tonnes.[597] The JSM were suggesting that NSW could build off-shore coal loading facilities capable of handling ships of 100,000 tonnes, and eventually 200,000 tonnes. The challenge was clear for the NSW Government, but it would prove to be wary of investing the major funds necessary in port and rail capacity the Japanese were looking for.

At times in 1970 a queue of ships stood off the port of Newcastle waiting to gain access to the coal terminals. This was only a temporary situation, but by late 1973, with Hunter Valley exporters signing new contracts for export to Japan and other markets, the port's capacity was under threat. And in its annual report for 1973-74, tabled in November 1974, the JCB would point to the failure of the NSW coal industry to meet its export contracts commitments, the failure being due to labour shortages and inadequate rail transport and ship loading facilities. The Queensland industry however would see massive expansion in the 1970s, largely unimpeded by transport constraints.

Queensland Government exits coal mining

The 1960s saw the end of the Queensland Government's operation of coal mines, a role it had played since its first foray into the industry in 1915. The record had been mixed, with the early failure of the Warra mine, the Government's purchase of the Mt Mulligan mine after the tragic explosion in the 1920s and the development of other mines, most notably the Bowen State Mine at Collinsville. The Bowen State Mine at Collinsville had a chequered history, which included a gas outburst in October 1954 which killed 7 workers. The Royal Commission appointed to inquire into the tragedy reported in February 1956 and was critical of many aspects of the mine's operation and management. The mine had been losing significant amounts of money, the main cause of these losses being what the Commission called "completely unsatisfactory production" – in other words operating inefficiently and well below its capacity. The causes of the inefficiency included excessive grades (the very steep grades of tunnels in the mine), poor roof conditions, faults in machinery and equipment, the uncooperative attitude of the union and the employees, and the shortcomings of management, particularly its lack of flexibility and initiative in planning and thinking.

The Commission found that the seven miners died from asphyxiation and were not injured by the flying stone and coal which had been hurled around 30 metres from the face. The outburst had been caused by an eruption of almost pure carbon dioxide which had been captured behind the face until it was able to escape as the miners were working at the face. The Commission noted that there had only been three other comparable outbursts in Australian coal mines, all at the Metropolitan mine in NSW, the first in 1897, the second in 1925, and the third just after the Collinsville outburst in December 1954. A wide range of measures to improve the efficiency and safety of the mine was recommended, but no blame was attributed to any individual for the outburst.

Mechanisation of the mine was one of the Commission's terms of reference, with the Commission concluding that the Minister's decision to mechanise the mine was correct. The Commission recommended that mechanisation of the mine should continue and that there was no basis for the mine to close. However the Commission warned that much of the task of ensuring the future success of the mine would fall on the shoulders of management. "If Management can find the secret of raising the morale of the employees it will have gone a long way towards overcoming production lags. We refuse to believe that this is by any means an impossible task. That the task is heavy is not disputed and therefore it seems to us that Management must be given all the aid possible."[598]

The possible sale of the mine as a going concern was canvassed in the Commission's report, but was not recommended as it believed that the Government would only be able to sell it for a "sacrificial price."[599] The favoured outcome for the Commission was for the Government to create a government-owned company to run the mine, with a board of directors comprising one or two directors with expertise in mechanised mining, one director with expertise in finance, one director with general management experience, and one director to represent the Mines Department or the Treasury. The Board, although responsible to the Minister for Mines, should be given the maximum degree of control, said the Commission.[600] However, if the mine is not profitable by June 1958, the Commission recommended a further review to determine its future. The recommendations in relation to the corporate structure of a new government owned company, which could be seen as well ahead of the time, no doubt owed much to the input of one of the Commissioners, Walter Scott (later Sir Walter Scott), who established Australia's first broad based management consultancy, W D Scott & Co, after World War Two.

Collinsville State Coal Mine continued on as a government-owned and operated mine, but by early 1961 troubles were looming. In February a dispute between the union and management over the interpretation of the regulations in relation to temperature saw 34 men dismissed for refusing to work in a section of the mine. The dispute was soon resolved, but in March the Queensland Collieries Employees Union was pushing for the closure of the open cut mines at Callide and Kianga to support the highest production possible at Collinsville. The QCEU president, T Millar, said that there could be more dismissals at Collinsville because of a fall in demand from Mt Isa and the Railways. Millar also warned that the nearby Bowen Consolidated open cut and underground mines at Scottville, about 4 kilometres from Collinsville, were stockpiling coal at a significant rate, implying that this was endangering the market prospects for Collinsville's coal.[601]

The OMS (output per manshift), a rough but useful indicator of productivity, achieved by Collinsville continued to be well below the average being achieved by other underground mines. In the second half of 1960 for example, Collinsville's OMS was only 2.36 tons compared with 3.3 tons for all underground mines. And in the first half of 1961, Collinsville's OMS fell to only 1.93 tons compared with 3.4 tons for all Queensland undergrounds. It was not surprising therefore that in April 1961 the Government brought down the axe, dismissing 208 men and closing the mine. It was then put up for sale and was bought by Dacon Collieries, a subsidiary of British company, Wood Hall. MIM would assume full control of Collinsville in 1977. Following the closure in 1961, the town of Collinsville was hit hard, as it was a one-industry town, largely reliant on the jobs in the mine. Collinsville remained closed until the second half of 1962 when it re-opened as Dacon No.3 or Bowen No.3.

In 1963, the remaining State coal mine at Ogmore, the Styx No.3 State Mine north of Rockhampton, was still operating, but was losing money. The Government announced that it was calling tenders for the purchase of the mine which employed 70 workers. The mine continued in production until the second half of 1964, when it finally closed, bringing to an end almost five decades of state-owned coal mines. However it would take NSW until 2002 before it moved out of the coal mining business.

New power stations approved

While the 1960s saw the end of the involvement of the Queensland Government in coal mining, the decade also saw the start of a major coal-fired expansion of the Government's electricity generation and distribution network. In 1962 the State Electricity Commission gave the go-ahead for the development of two new power stations, one at Swanbank near Ipswich, and the other near Biloela on the Callide coalfield. The Swanbank A station, with 6 units of 66MW was commissioned in 1967, with the Swanbank B station (4 units of 120MW capacity) commissioned in 1971. Callide A, with 4 units of 30MW capacity was commissioned between 1965 and 1969. The new Swanbank station saw 3 local collieries participating in the supply of 20,000 tons of coal per month. In 1965 approval was given for another new station at Collinsville with 4 units of 30MW capacity; this station was commissioned in 1968, with a 60MW upgrade commissioned in 1976.

These new power stations were needed to supply the State's growing population and mining and industrial sectors and while modest in size compared with later power stations such as Gladstone and Tarong, gave a boost to the coal mining industry. By 1970, the West Moreton's production had hit a record 2.1 million tons (around 1.9 million tonnes), double the level twenty years earlier. Callide production had reached almost 400,000 tons and the Bowen district almost 500,000 tons by 1970. However while production from the West Moreton mines would continue to grow, major concerns would emerge in the 1970s over the future of the underground sector in that district. The new Gladstone power station (4 by 275MW) would start to produce power in 1976 and a new transmission line to Brisbane would mean competition for Swanbank to supply the South east of the State; a number of West Moreton mines would close, and only 11 undergrounds would be operating by the end of the 1970s.

In NSW, the 1950s had seen three new coal-fired power stations developed – Wangi on Lake Macquarie, Tallawarra on Lake Illawarra and Wallerawang near Lithgow. New and larger stations followed in the 1960s, with Vales Point (3 by 200MW and 1 by 275MW) on Lake Macquarie commissioned between 1963 and 1966, and Munmorah (4 by 350MW) on Lake Munmorah, part of the Tuggerah Lakes, commissioned between 1967 and 1969. The development of the big Liddell Power station (2000MW) was announced in 1964, with the first unit due to begin generating in 1971. As in Queensland, these new stations were a positive for the coal industry. Three new collieries were developed to supply Vales Point - Chain Valley, Newvale No.1 and Wyee State. Munmorah State and Newvale No.2 collieries were developed to supply the Munmorah station.

Coal and Allied Industries is born

One of the major developments of the 1960s was the merging of the JABAS and Caledonian groups in 1960 to form Coal and Allied Industries Limited. The new Coal and Allied group was now the largest producer in the northern district, with 13 mines, 10 of these in the Cessnock district. Caledonian's roots went back to its formation in Scotland in 1895 and its purchase of major NSW collieries in the 1890s and early 1900s, with Howard Smith becoming the major shareholder by 1912 when it formed Caledonian Collieries Limited, a locally registered company. As we saw earlier, JABAS was formed in 1931 with the merger of J&A Brown and Abermain Seaham, and J&A Brown's roots extended back to the Brown family operations in the 1840s. Howard Smith emerged from the 1960 merger retaining a 30%

interest in Coal and Allied and would remain as a major shareholder until a major restructuring of the company in the 1980s.

Being heavily dependent on the Greta seam mines in the Cessnock district, Coal and Allied's future in the early 1960s was somewhat uncertain, with the local markets for Greta seam coal continuing to face massive competition. Christopher Jay described the merger of the JABAS and Caledonian groups as "a merger facing incipient disaster", although subsequent developments in the market would see Coal and Allied continuing to be a strong force in the industry.[602] The markets for Greta seam coal, particularly the gas companies and the railways, had changed significantly in the 1950s, but in the 1960s the discovery of natural gas in the Cooper Basin in SA, in Bass Strait and in the Roma area in Queensland would cause even more disruption. The first gas field in Bass Strait was discovered in 1965, and was followed by other major discoveries. The Moomba field in the Cooper Basin was discovered in 1966, with the first gas supplied to Adelaide in 1969, and with the pipeline to Wilton on the outskirts of Sydney completed by December 1976. Gas was discovered in the Roma area in Queensland in 1900, but it was not until 1969 that the first sales were made to the Brisbane market.

Natural gas was not only a looming problem for the coal industry, which was continuing to suffer from competition from cheap fuel oil. Edward Warren, chairman of the Australian Coal Association, called for action by the Federal Government to counter what the industry saw as unfair competition from the oil refineries. The survival of the industry was at stake, according to Warren, with the oil refineries having "not the slightest intention of doing anything positive to modify the unreasonably high furnace oil (ie fuel oil) output from their refineries."[603] Warren said that the result was that "coal is being ousted from home markets where oil has no intrinsic advantage" only because of what he said was short term price competition. Warren added that the commissioning of two new refineries in the Brisbane area was likely to exacerbate the problem. There was no action by the Federal Government to assist the coal industry to compete with fuel oil. With cheap fuel oil already disrupting the local markets for coal, and natural gas looming as another major problem, the outlook for the Greta seam producers and their employees was grim.

Production from the main Greta seam peaked at around 3.2 million tons in 1954, but had fallen to 2.8 million tons by 1960. By 1964 production was down to 2 million tons, but did start to recover over the next few years, reaching 2.5 million tons in 1967, thanks to the growth in the export market. Mine closures were common in the 1950s and continued into the early 1960s, with the Greta seam mines accounting for many of these. In the 1960-61 year alone, 11 mines closed in NSW, 6 in the South Maitland coalfield, and only 11 mines were left operating on that field by the end of the year. The JCB, well aware of the prospects for some mines and their age profile, warned that more old mines were likely to close.[604]

Coal and Allied's ageing operations were among the prime candidates for closure in the early 60s. Pelaw Main, whose history dated back to 1902, closed in 1961. Pelaw Main was developed from the early 1900s out of the old Stanford Greta No.2 Tunnel mine acquired by J&A Brown in the 1890s. In 1961 Stanford Main No. 2 ceased operations, with 175 dismissed in June, and the remaining 70 losing their jobs once the equipment had been removed. Many of that mine's workers lived in the local town of Paxton, population 585, and there were fears that the mine's closure would threaten the ongoing viability of the town. At the Aberdare group of mines which traced their history back to 1901, 1961 also saw major retrenchments – 69 from Aberdare No. 2, and 50 from Aberdare No.7. Aberdare Central also closed in 1961, with 50 miners dismissed and with the remaining 20 losing their jobs after equipment was removed from the mine.[605] The famous Richmond Main colliery was finally closed in 1967. Richmond Main, developed from 1912 by J&A Brown from the old Richmond Vale Estate purchased in 1897, had produced over 14 million tons of Greta coal over its lifetime, but had reached the end of its economic life.[606]

But Coal and Allied also re-opened Caledonian's old Aberdare North in 1965, which continued operating until 1988, and it developed a new colliery, Aberdare East, which was given approval by the JCB to open in 1962 to supply the growing Japanese market. The Miners' Federation expressed concerns that Aberdare East would be able to produce coal more cheaply than existing mines and pushed the NSW Government to impose a production quota.[607] However with Aberdare East planning to employ 156 workers, the JCB supported the development of the new mine which, it said, would take another 12-18 months to complete, with its opening to synchronise with the closure of another mine. The JCB also noted that the new mine would support a continuing mining base in the Cessnock area. Aberdare East in fact took longer to develop, with major development works only commencing in 1966.

The consolidation of the industry during the 1960s saw the major coal producers accounting for just over 28 million tons of production, or 81% of the NSW total by 1969-70. BHP (including its AI&S subsidiary) was the largest producer with an output of 7 million tons, followed by Clutha with 6.9 million tons, the two NSW Government groups of mines (operated by the Electricity Commission and State Mines Control Authority) at 6.8 million tons, and Coal and Allied with 5.1 million tons. The two other major companies were KCC and R W Miller with 1.2 million tons each.

Fortunately for Coal and Allied, in the early 1960s the Japanese market for gas making coal was expanding. The major gas companies in Japan used coal to make gas for local household and industrial consumers and imported a significant quantity of their requirements. The Coal and Allied Greta seam mines in NSW were well placed to supply companies including Osaka Gas Company, which was already a customer of Caledonian, and Tokyo Gas Company, which had received its first shipment of Caledonian coal in 1960.[608] The Coal and Allied mines were claimed to be the only ones in the north capable of supplying the quality gas coal, with less than 5% ash, demanded by the Japanese. The company also owned 3 mines which were not mining the Greta seam – Liddell in the Upper Hunter Valley, which produced soft coking coal, Wallarah, on the eastern side of Lake Macquarie producing steaming coal for the Sydney market, and Stockrington, near Minmi, which also produced steaming coal.

And by the mid-1960s, with the Japanese steel industry looking for even larger coking coal supplies from Australia, the prospects for Coal and Allied were improving. In 1965 Sir Edward Warren secured a contract for Coal and Allied with the JSM for 3 million tons of Liddell coal to be supplied over 5 years commencing the following year. Warren said that he expected the company's total exports to Japan over the next 5 years would be between 4.5 and 6 million tons per year. Warren also said that the new contract was evidence of the importance of the construction of the new 2,000 tons per hour Basin coal loader in Newcastle.[609] By 1968, the company's exports to Japan were running at over 2 million tons per year and the following year new contracts with the Japanese were negotiated and existing contracts revised; the company was now looking at exports of over 21 million tons over the coming 5 years and beyond, with improved prices also secured.[610] The hunger for coal from the expanding Japanese steel industry during the 1960s was a godsend for the northern district, and would expand further in the 1970s, although exports from NSW to the JSM would never reach the scale of the Queensland industry with its massive deposits of high quality hard coking coal mined from huge open cut operations.

US billionaire bursts onto the scene

US billionaire Daniel Ludwig made a dramatic entry into the NSW coal industry in 1965. His Clutha Development Pty Ltd bought Placer's coal interests in an open cut mine in the Hunter Valley and mines in the Burragorang Valley south west of Sydney which had been purchased from S&M Fox, and also bought the coal interests of CRA in the Burragorang Valley which CRA had mainly acquired from the Clinton group. Although Ludwig was not directly involved in the Australian coal operations, his name and fame attracted the media's interest. He was one of America's wealthiest men, reputed at one stage by Time Magazine to have had a personal net worth of over $2 billion, a massive sum for those days. Ludwig ran the largest fleet of ships in the world, and his extensive business interest also included hotels, a coal mine in the US state of Virginia and petroleum refining.[611] He had purchased the coal mine in Virginia with a view to the Japanese market, but the Japanese had reduced their imports. Clearly his entry into the Australian industry was with a view to the now expanding Japanese market.

Ludwig's Clutha coal interests by the end of the 1960s included the Oakdale State mine which it had bought from the NSW Government in 1968 (outbidding a JASBAS/ Newcastle Wallsend Coal consortium and an Ampol/ Kathleen Investments consortium), the newly developed Brimstone 1 and Brimstone 2 mines, and the mines purchased in 1965 (Wollondilly and Wollondilly Extended, Nattai-Bulli and Valley). Clutha was now a major force in the NSW industry. In the late 1960s, with exports to Japan growing, the limited capacity of Port Kembla was a major issue for the industry, and in particular for the Clutha group which exported its Burragorang Valley coal through Port Kembla. Exporters began to discuss options with the NSW Government which had not been keen to invest its own funds in new infrastructure for the coal industry, particularly as the previous Labor Government had spent $10 million on the new loader at Port Kembla and which could be expanded in capacity although that would be insufficient to meet the projected needs of the southern exporters. One of the Clutha suggestions, from George Jennings, its chief general manager, was for construction of a pipeline from the Burragorang Valley to a new offshore loader. Mines Minister Wal Fife did not accept the pipeline idea and instead suggested a new rail line.[612]

Clutha then came up with what became one of the most controversial proposals ever seen in the NSW coal industry. It proposed to build its own rail line from the Burragorang Valley to the edge of the escarpment, from where it would run conveyors to a new offshore loader at Coalcliff. Despite the controversy this proposal created, the NSW Government approved the Clutha proposal in November 1970, obviously with an eye to the revenue it thought it would make. NSW Premier Askin's announcement of the Government's approval referred to an expected revenue stream for the Government of at least $100 million over twenty years from the company, based on at least 7 million tonnes of coal a year being moved down to the coast.[613] The Government passed the Clutha Development Pty Ltd Agreement Act in 1970 to authorise the construction of the railway and offshore loader. Similar Acts were common in the Queensland industry in the 1960s and 1970s, but this was unusual to say the least for NSW. By 1971, the Clutha proposal was looking doubtful and the company was reported to be reviewing the project; the project was shelved by the following year.[614]

By 1972 Clutha had become a takeover target, attracting the interest of BHP which had agreed to buy the company on the condition that the NSW Government approve BHP constructing and operating its own private rail line to O'Brien's Drift on the escarpment west of Wollongong. Negotiations between BHP and Clutha however broke down in 1972 when the NSW Government and BHP failed to reach agreement on the BHP's proposal and the Government's counter proposals. The Government rejected BHP's proposal and countered with a proposal that BHP would pay a security deposit with the Government to cover the cost of the rail line, the purchase of rolling stock and other expenditure; the Government would own the line and repay the deposit from revenue it earned from the rail freights for the coal. BHP rejected that proposal despite the Government twice reducing the rail freight rate it proposed to charge. Once it was clear that BHP had rejected the Government's proposal, the Government came back with an alternative plan whereby BHP would build the rail line while the Government would provide the rolling stock and also the crews to operate the trains; BHP would pay the agreed freight rate plus a franchise fee per ton for all coal carried on the line. However, for BHP the economics of the Government's proposals did not stack up, the takeover did not proceed and Clutha continued as a privately owned company.[615] Had BHP been able to purchase Clutha it would have become the largest coal miner in Australia, at least until Utah emerged in the 1970s as the dominant producer in Queensland.

Clutha was also pursuing a new development at Sirius Creek in Queensland in the late 1960s; the underground project was around 30 kilometres south of Blackwater with a planned output of 5 million tonnes per year for export to Japan. The company's proposal included a new rail link to Sabina Point, south of Mackay, and a 20 million tonnes per year coal loading terminal on nearby Akers Island capable of accommodating bulk ships of up to 200 DWT. Trial mining proceeded in the early 1970s, but an explosion in one of the mines saw it closed and operated on a care and maintenance basis during the 1970s. Given the Queensland Government's stranglehold on the railway system and the revenue it was earning from its rail system, it is doubtful that Clutha would have been successful in securing Government approval to develop its own rail link to the coast had it wished to proceed to full development of Sirius Creek.

NSW industry starts a reorientation to the export market

By the start of the1960s, the NSW coal industry was more optimistic about its prospects, and this was reflected for example in the JCB's annual report for the year 1959-60. The JCB saw a bright future for the NSW coal industry, reporting that the industry "faced the next decade with great confidence." It expected exports from NSW to Japan to grow to 3 million tons a year by the mid-1960s, with the southern Bulli seam coking coal exporters reaping most of the benefits. Soft coking coal exports from the northern district would also be a beneficiary, with around 750,000 tons of annual exports likely.[616] The late 1950s to early 1960s was the period when the NSW coal industry once again began to become a serious player in the export market. Following the downturn in the NSW industry in the late 1920s, and in particular the northern lockout in 1929-30, the overseas markets for NSW coal largely disappeared. Exports were negligible in the 1940s and for most of the 1950s, but thanks to the growing Japanese economy, its booming steel industry and the decline in Japan's domestic coal industry, the opportunities were now opening up for Australian producers.

The introduction of a new plan to restructure the Japanese coal mining industry in 1963 caused some major concerns in Australia, for a short time threatening the soft coking coal exports through Newcastle. The Japanese plan was to see coal production increase over the following years, and the Government was also looking to the steel mills to increase their purchases of soft coking coal from local Japanese mines. The Japanese steel mills however wanted to buy the cheaper imported soft coking coal. There was concern that our exports to Japan would fall away just at a time when the contract for construction of the new coal loader in Newcastle was still to be finalised. Japan was also experiencing economic pressures in the early

1960s, including with its balance of payments, and at one stage directed the steel mills to cut production by 30%.[617] The JCB looked to the Australian Government to make representations to the Japanese Government, fearing that "unless the coal loader was installed, the chances of being able to sustain Australia's competitive advantages in the northern coalfields would be low", adding that "If the NSW Government pulled out of the project.. it would be difficult to revive it later."[618] Following representations by the Australian Ambassador in Tokyo and the visit by NSW Premier Heffron to Japan in April 1963, the concerns of the JCB and others fortunately did not eventuate, with the chairman of the Japanese Iron and Steel Federation, Shigeo Nagano, promising Heffron that Japan would obtain all of its soft coking coal imports from Australia. The Japanese steel industry grew rapidly during the 1960s and by 1970 its production had reached 93 million tonnes, or 15.7% of global production, compared with only around 6% in 1960.

However by the mid-1960s, with NSW coal exports exceeding 6 million tons, there were concerns over the rapid growth in coking coal exports and the impact on limited coal reserves. Over 90% of those exports were coking coal, mostly from the southern coalfield, and the JCB noted that the growing demand was forcing the southern mines to extend their operations much deeper to over 300 metres underground. In its annual report for 1955-56 released in October 1956, the JCB was extremely optimistic about the potential for the Australian steel industry to grow and said that local demand for coking coal would be five times the then current level by the year 2000, with the 2000 figure expected to reach 30 million tons. But the Board cautioned that "Whether Australia has sufficient cheap coking coal to encourage the growth of the Australian iron and steel industry on the scale that the market would otherwise permit, and at the same time permit unrestricted export, is a matter which is causing concern for the Board at the moment."[619] With the Bowen Basin development underway in the mid-1960s and then to surge in the 1970s, the JCB's concerns soon proved to be misplaced. The major problem for the NSW coking coal producers would be their ability to compete with the new Queensland mines.

Wonthaggi colliery finally closes

The rich and troubled history of the Australian coal industry of course was not confined just to NSW and Queensland, and any discussion of the 1960s would be lacking if mention was not made of the Wonthaggi mine in Victoria. As noted earlier, Wonthaggi was developed by the Victorian Government from 1909 to supply coal to the Victorian railways, at a time when industrial disputes in NSW were causing major problems for interstate customers. The mine, situated in the west Gippsland region around 8 kilometres north of Cape Paterson and exploiting the Powlett River Valley seams, was the largest black coal mining operation in Victoria. The operation was managed by the Victorian Railways and in fact covered a number of mines, shafts and tunnels; by 1919 there were 8 mines in production. The seams were thin and the mine had a poor record for safety. Coal seams at Cape Paterson had been discovered by the explorer William Hovell in 1826 and then "rediscovered" by Samuel Anderson, a local resident, in 1838. Some limited mining at Cape Paterson did occur in the late 1850s and early 1860s but it was not until Wonthaggi was developed that we saw a significant coal mining operation in Victoria.

A rail link to Melbourne was completed in 1910 and the development of the town of Wonthaggi commenced, with 100 cottages for miners built by the middle of that year. During 1929-30, the first year of the Great Depression, employment at Wonthaggi averaged 1776 and production was 662,000 tons, that year marking the peak for both employment and production. During the 1930s the mine became best known for its turbulent industrial relations, culminating in a 5 month strike in 1934, and for a tragic explosion in 1937. The explosion occurred in one of the shafts and killed 13 maintenance workers who had gone underground that morning. At the time of the explosion, a stop work meeting was being held in the town to protest at the safety problems at the mine. Idris Williams, the general president of the Miners' Federation from 1947, had made his mark as vice president of the union's Victorian branch during the 1934 Wonthaggi strike, and then as State president. Williams, a communist, was a key factor in Wonthaggi gaining the title of "Red Wonthaggi", but was a popular local figure and was honoured as the town's "finest citizen" and "favourite son".[620] During the 1949 national strike, Williams was imprisoned by the Chifley Government along with several other Federation officials.

The Victorian Government finally closed the mine at the end of 1968, by which time almost 17 million tons of coal had been produced. Wonthaggi's closure marked the end of the last significant black coal mine in Victoria, although the town of Wonthaggi lives on as an important centre for the local area. By 1968, employment at the mine was down to around 100 and the market for its coal was rapidly disappearing as diesel locomotives took over from the old steam powered locomotives.

One of the features of the Wonthaggi mine when it closed was the need to find a home for its 104 pit ponies, the horses used underground for decades for much of the heavy work. An article in the Australian Women's Weekly marking the closure noted the public interest in the future of those ponies. But the article also quoted the miners as asking why there was so much fuss about the ponies and not about their own future, with many of the miners who were losing their jobs having worked at the mine all or much of their working life.[621] The mine's manager, Jim Byrne, told the Weekly that he had received hundreds of letters and phone calls from the public inquiring about the ponies and offering to adopt them. Some of the callers were offering to look after the ponies in return for "light duties", while others said there would be no strings attached. The mine's boss wheeler, Ray Williams, the man who trained the ponies for the wheelers who managed the ponies underground, said that people thought of the ponies as kids' ponies. "They're not" he said. "They are horses which have been working underground with the men." The Weekly noted that the ponies had worked underground nearly all their lives, and most of them had never been ridden. They were not used to reins, they worked by voice, and they were too old to learn new jobs, the article said.

Wonthaggi was not the last coal mine in Australia to use pit ponies. That record was held by the Collinsville mine in Queensland and the pit ponies would feature in a unique and somewhat comical industrial dispute which culminated in 1990 the retrenchment of the last ponies that had been enrolled as members of CFMEU.

Miners' Federation open cut ban lifted

International aluminium producers in the 1960s were taking a closer interest in Australia as a place to invest in new smelters to convert alumina into aluminium metal. We had the bauxite (the basic raw material which when refined becomes alumina), the land and port facilities, the coal for electricity, and critically our power prices were reasonable, although not cheap enough in NSW to tempt the international aluminium companies. Aluminium has often been called solid electricity because of the huge amounts of electrical energy required to produce the metal. The Bell Bay smelter in Tasmania, built to draw on the state's cheap hydro power, came on stream in 1955. Bell Bay was built as a joint venture between the Tasmanian and Commonwealth governments, and was purchased by Comalco in 1960. In 1963 Alcoa's Point Henry smelter near Geelong in Victoria began production, and the company's alumina refinery in Kwinana WA began production in 1964.

In the 1960s, US consultants Ebasco, were engaged by the NSW Electricity Commission, the State Electricity Commission of Victoria and the Snowy Mountains Hydroelectric Authority to advise on what additional capacity would be needed by the two states. Ebasco told the NSW Commission that they saw value in the open cut coal resources of the Upper Hunter Valley for use in new power stations. Ebasco recommended that a new station be built at Ravensworth, near the future Liddell station around 1966. However, with the construction of the Munmorah station proceeding, the NSW Government decided to defer the construction of a new Hunter Valley station. The Government also understood the Miners Federation's bitter opposition to such a plant unless it drew its coal supplies from underground coal mines.[622]

The NSW Electricity Commission was negotiating with Alcan, the major Canadian aluminium producer, in the mid-1960s to build a smelter in NSW. An attractive place to invest was the Hunter Valley, with suitable land, the port of Newcastle nearby and good coal resources to produce electricity at a reasonable price. There was a major problem – the Electricity Commission would need to build a new power station to supply the necessary electricity, and for the electricity price to be competitive, access to cheap coal was vital. The Commission had plans for a new power station at Liddell near Muswellbrook in the Upper Hunter Valley which could supply the power, but with the union ban on new open cut mines, it had no obvious source of cheap coal.

The Commission' chairman, Aubrey Coady did a "brilliant negotiation" with the Miners' Federation. In return for lifting its opposition to new open cut mines to supply the Liddell power station, the Commission would undertake to locate the smelter near Kurri Kurri, in the area where major job losses in coal mining had been concentrated.[623] Coady "secured the miners' acceptance of the Liddell open cut (mine) as the only way NSW could achieve power prices low enough to attract an aluminium smelter to Kurri Kurri….That agreement … was to have a dramatic impact on the cost of production of electricity in NSW and would in time allow the development of the even larger Bayswater power station" built virtually on top of the coal resources in the Upper Hunter which could be mined by open cut methods.

NSW Premier Jack Renshaw had made the first announcement about the Liddell power station in October 1964, saying that construction would commence immediately, with the first unit to begin operating in 1971, and with coal consumption to reach 5 million tonnes per year.[624] The Government, he said, wanted to attract heavy industries to the Hunter Valley and the new station was a key part of that plan. Labor MPs from the northern coalfields were critical of the choice of Liddell for the new power station, arguing that it would involve significant new open cut mining, contrary to the policy of the JCB which was for open cut mining to operate in conjunction with underground mining. The MPS also said there was a more suitable site at Broke where land was available. Mines Minister Pat Hills said that he was aware of the concern that the announcement would generate, but stressed that Liddell involved a new outlook on coal production; Hills also claimed that he had given a clear indication that all the workers employed in the underground sector would continue to be employed.[625]

Labor lost power in May 1965, and in December Alcan and the new Liberal Premier Robert Askin announced that an aluminium smelter would be built in the Newcastle – Port Stephens area. Askin said that the construction of the smelter would commence in 1967, and the smelter would be operational in 1969. Direct employment was estimated at 400, with another 1500 jobs said to be created in service industries. Alumina was to be supplied from a new plant being constructed in Gladstone, in which Alcan had an equity interest. Alcan said that the NSW Electricity Commission had contracted to supply the substantial power that would be required.[626] In 1966, Alcan announced that its smelter would be located in the Cessnock district, with power to be supplied from the Liddell power station.[627] Following a call for tenders, the Electricity Commission announced in December 1967 that coal supply contracts had been awarded to Costain (50 million tonnes over 14 years) from a new mine at Ravensworth, commencing in 1972, and to a consortium in which the major companies were Coal & Allied and Wood Hall (17.5 million tonnes over 14 years) from a new mine at Swamp Creek, commencing in 1970). The leases for both new mines were held by the Commission.

The Miners' Federation, according to long time official Edgar Ross, had tried to convince the NSW Government to become involved in mining the coal for the Liddell power station. "All attempts by the Federation failed…" he said "….as did the efforts to get the Government to honour the undertaking of its Labor predecessor that the coal would be won by underground and not open cut methods."[628] However the formal reaction of the Miners' Federation to the new coal contracts for Liddell was relatively muted. Common Cause reported that the contracts had "led to protests from mining unions at the decision of the Askin Government to depart from the established policies of getting coal from State mines and by the method of underground mining."[629] The Federation's Central Council passed a resolution supporting the stance of the union's Board of Management "in opposing the supply of coal to the new Liddell power station exclusively by open cut mining…. Further we see the decision to contract out the supply of coal to the station to private interests …is a change of policy by the (Electricity Commission) …Past policy of the (Commission) namely the siting of power stations on coalfields and supplying coal by mines owned by Government instrumentalities has been in the best public interests."[630] What was not mentioned by the Central Council was that, with Liddell sited in the Upper Hunter Valley near Muswellbrook, supply of coal by open cut mines was the logical option and was the option which a Labor Government would no doubt have pursued had it continued in office after 1965.

The new Liddell power station and the new open cut mines developed to supply it were certainly landmark events in the Hunter Valley, and set the stage for other major new open cut developments in the 1970s and 1980s. In 1979 the NSW Government announced that another new power station at Bayswater near Muswellbrook would commence operating in 1985, with coal to be supplied by Costain's Ravensworth mine and the Liddell State Mine.[631]   However the Federation's antipathy towards open cut mines would come back with a vengeance in the early 1980s, when the development of the Drayton open cut would be the subject of a ban by the Federation and would create a split within the Federation's membership.

The Joint Coal Board too had to bend its own policies to approve the new mines for the Liddell power station. The JCB's balanced development policy regarding underground and open cut mining involved companies wherever possible operating both types of mines: "The Board believes that as a general rule in designing colliery holdings any available open cut coal should be included with a larger proportion of underground coal to enable a balanced programme of production to be maintained with benefit to the average cost of the combine operation."[632] In other words, if companies wanted to mine the cheaper open cut coal, they generally would need to also run higher cost underground operations. The policy was essentially designed to protect employment in the underground sector and maintain the skilled underground workers in the industry. The boilers of new Liddell power station however were designed to operate using the lower grade, high ash coal which the Ravensworth and Swamp Creek mines would supply. The JCB simply made an exception to its policy for these mines, a decision which reflected the reality of the needs of the next generation of power stations which had to buy cheap coal in order to supply competitive electricity to the aluminium smelters or other major industries.

Continuous miners spread; early longwall trials fail

The mechanisation process which began in earnest in the 1950s continued into the 1960s, with the adoption of continuous miners a major feature of that decade. In 1960, the industry in NSW had moved a fair way down the mechanisation path, with almost 90% of underground coal produced cut by mechanical equipment, and 92% of coal loaded mechanically. However the spread of continuous miners was about to occur, and this would produce significant increases in the standard measure of productivity – output per man shift (OMS). OMS in the underground sector in NSW was 2.8 tons in 1950, rising to 3.1 tons in 1955 and to 4.3 tons in 1960. However by 1970 OMS more than doubled to 10.3 tons, and by the early 1980s it was still at that level, with further significant improvement dependent on the spread of longwall mining during the 1980s and 1990s. As we saw in chapter 8, the first continuous mining machines were introduced into the JCB and BHP mines in the early 1950s. By 1960 there were 71 continuous miners in use in NSW, but this would grow to 191 by 1970, when these machines accounted for 87% of daily coal production.

Since the 1980s longwall retreat mining has become the most common form of underground mining in Australian coal mines, but it is a relatively recent innovation. Longwall mining involves the driving of two roadways at right angles to the coal face which may be between 100 and 300 or more metres wide; these roadways form the sides of the block of coal to be mined. Another roadway is driven at right angles to connect these two roadways and to expose the longwall face (the coal face) for mining. A coal cutting machine (a shearer, which is usually electrically powered) is installed which travels along the coal face shearing off the coal, which is then transferred to a conveyor. A series of hydraulically powered supports holds up the roof of the mine. These supports are then moved forward to allow the shearer to take its next pass along the coal face. The roof of the mine in the void behind the supports (known as the goaf) is then allowed to collapse. The whole operation continues to retreat away from the mined section until the rectangular block of coal is extracted. The longwall system is then re-located to the next block to be mined.

The first attempts to introduce modern longwall mining in NSW were on the southern coalfields, at the Coalcliff colliery at the end of 1964. Coalcliff however found that there were major problems with the equipment, the weight of the roof and with the volume of dust produced and soon abandoned the method. There was a second attempt at operating a longwall in 1965 but this also was unsuccessful, with the company removing the equipment and storing it for possible future use. The South Bulli and Kemira collieries were also operating longwalls by 1965. South Bulli experienced similar equipment and dust problems to Coalcliff and also discontinued the use of the longwall. BHP's Kemira colliery however was reported as "achieving remarkable success, reaching production figures considered possible but not probable."[633] South Bulli persisted with its longwall, although it continued to encounter technical problems. Kemira's experience also proved to be problematic, and in 1967 it decided to remove its longwall. It would not be until the late 1970s that longwalls proved their worth and viability, and by the 1990s, the longwall system would be widespread throughout NSW and Queensland, with new underground mines generally designed from their initial development to incorporate this technology.

In 1968, South Bulli colliery was able to complete the mining of coal from its longwall panel in less than 6 months, compared with a 2 year period for its first panel which was of a similar dimension. This and other achievements at Kemira colliery were giving the industry confidence that longwall mining was now a viable technology. BHP's John Darling colliery was preparing to operate its new longwall in 1969, and the NSW Mines Department was positive in its annual report, saying that "The extension of this form of mining into other mines and districts (is considered) only a matter of time" and would be welcomed for the greater safety it would provide.[634] In 1970 for example, in mines with these longwall systems, and in mines using shortwall systems (which both involved powered hydraulic supports), there was only one accident caused by a fall of coal or stone from the roof or sides.[635]

The new longwall technology also had a major impact on the working arrangements in the industry. In 1957 BHP applied to the CIT to allow coal production on the afternoon and night shifts, but its application was rejected, with the CIT recognising the long standing custom in the industry that there should be no production on these shifts. However the CIT ruling allowed that production could occur where this was agreed, or where it was regarded as essential following an investigation by an industrial authority, and in relation to BHP's Elrington colliery (the subject of the BHP application) it did allow production from pillars "to avoid the dangers of coal creep, heating and seepage."[636] It then took another ten years, but in 1967 the ability of producers to operate their mines more intensively was given a boost when the CIT granted an application by AI&S, supported by the NSW Colliery Proprietors' Association and the JCB, to allow 24 hour production on a longwall face. And in March 1970, the CIT granted an application by AI&S for 24 hour operations to be extended to operations which supported the longwall operations including development work to prepare new areas for longwall mining.

With the unions fiercely opposed to the extension of hours and with major strike action continuing in the industry, in July 1970 Gallagher decided to pursue the question of shorter hours which were being sought by the unions. At the end of July Gallagher awarded mine workers in NSW, Queensland and Tasmania a standard 35 hour week commencing in July 1971 (with five shifts of 7 hours per week), with a 37.5 hour week to apply from August 1970 (five shifts of 7.5 hours). Mine workers employed by the NSW Electricity Commission and the State Mines Control Authority were already working a 37.5 hour week, and were awarded a 35 hour week commencing in August 1970. In his decision, Gallagher referred to one of the International Labour Organisation's principles that the ordinary working hours for miners should be lower than the hours worked in industries in general. As part of the same decision in July 1970, the CIT also gave the employers the right to operate to produce coal 24 hours a day from Monday to Friday. While the earlier decisions were important, this decision marked the real start of a movement to more flexible working hours and arrangements which would be vital if the industry was to invest in major new technologies in both the underground and open cut sectors.

Of course mechanisation in the coal industry was not confined to continuous miners and longwall mining; it involved many aspects of the operation of a mine. The 1950s, 1960s and subsequent years saw changes in areas such as roof support technology, coal cutting, underground loading and haulage, and ventilation. Roof bolting was a major advance in underground mining in the post war years, with the first experiments in NSW at BHP's Elrington mine near Cessnock in 1949. Roof bolts are steel rods up to almost 2 metres in length which are drilled into the roof areas of a mine and which employ a resin to anchor the bolt to the rock or strata above. The early bolts used some sort of mechanical anchor. Over time, the use of roof bolts did away with the need for timber props and so opened up roadways and enabled existing mines to employ continuous miners and shuttle cars.

The early use of roof bolts in NSW was not met with a great deal of enthusiasm by the NSW Mines Department. Its annual report in 1951 for example said that the use of bolts increased that year, but the results did not indicate that this would replace conventional timbering. The Department also said that several mine managers were relying too much on roof bolting, and warned that roof bolting should only be used in conjunction with normal timbering. Departmental inspectors had been advised to take this issue up with colliery managers.[637] But later in the decade, roof bolting was starting to become accepted as evidenced by BHP's Burwood Colliery which the Mines Department described in 1958 as fully mechanised, and employing roof bolts in its operations in the Dudley seam.[638] Corrimal colliery on the south coast was another mine beginning to use roof bolts that year, and AI&S's Nebo colliery south of Wollongong was installing roof bolts in all of its development work, together with steel and timber supports due to difficult roof conditions.

Roof bolts and continuous miners were major factors in reducing deaths and serious injuries as, not only was the roof more secure and less likely to see large sections fall, but the use of continuous miners meant that miners working at the coal face were in fact a little way back from the face. Continuous miners also had hydraulic jacks which could help support the roof.

The introduction of diesel powered machinery underground was another major advance in the industry, with diesel shuttle cars being used from the mid-1960s. Shuttle cars have been described as the "workhorses" of the underground industry; they received coal from the continuous miner and then transported it to an area in the mine where it is typically loaded onto a conveyor. Diesel powered equipment was already in use in Europe in the 1950s and was approved for use in underground mining in the USA in 1954 following detailed testing by the Bureau of Mines; prior to 1954 the use of diesel machinery had not been possible in the USA because of regulatory restrictions on machinery which operated on combustible fuels. The US Bureau however had recognised that it was important to eliminate hazards associated with underground haulage systems operating from bare trolley wires and conducted investigations on the practicability of underground diesel haulage. Petrol engines had been discouraged underground, but diesel was able to be used much more safely than petrol; the fuel was safer to handle and most important, diesel engines did not require an electrical ignition system and so did not have the potential to create sparks which could cause fires or explosions. Diesel powered machinery also did away with the need for many of the potentially dangerous electric cables underground.[639] The first diesel powered shuttle car was introduced into service in AI&S' Appin colliery in 1965, and according to the Mines Department had proved to be extremely reliable.[640] Another early adopter was the Munmorah State Coal Mine on the NSW Central Coast, which was being developed to supply the new Vales Point power station; it introduced two diesel shuttle cars in 1966.[641]

In NSW the Mines Department was cautious about the introduction of the new diesel equipment, although it did understand the potential

safety benefits, saying that the new shuttle cars would overcome
the problem of cattle fires which had plagued the industry for years.
However it also warned that a higher standard of ventilation was needed
to prevent the build-up of toxic diesel gases. The same year also saw
the introduction of diesel powered personnel cars used for transporting
workers underground, and the approval for the use of a diesel powered
scoop for loading coal at the coal face, but the Department also warned
that the continued use of the personnel cars and the scoop would depend
on the operators complying with the regulations; failing compliance,
the Department said that approval to use these machines would be
withdrawn.[642] However just two years later, the Department reported
that a record number of diesel powered machines had been introduced
into the industry and these machines had played an important part in
improving efficiency and safety in the industry.[643]

Queensland underground mechanises; many mines close

In the 1950s, mechanisation in Queensland proceeded slowly, held back
by a range of factors. But that was about to change. By 1960, less than
14% of the coal mined underground was produced by fully mechanised
mines. By 1965, almost 75% of coal from the underground sector was
from fully mechanised mines, and by 1970 the percentage had grown to
95%. And the fully mechanised mines by 1970 were achieving an OMS
of 9 tons, almost double the level of a decade earlier.[644] However in 1970
there were still a surprising number of very small underground mines;
one of these, the Rosewood No.2 colliery in the West Moreton, the last
completely unmechanised mine in the State, produced a grand total of
12,590 tons of coal in 1969-70. There were only 6 collieries in the West
Moreton and 2 in the Bowen district which produced over 100,000 tons.

As we saw in the last chapter, the QCB's annual report for 1959-60
introduced for the first time data on the extent to which the industry was
being mechanised, with production and productivity (OMS) categorised
by the degree of mechanisation. What the data did not show however, were
the numbers of mines in each category. That information became clear

in the QCB's annual report for 1961-62, when the Board reported that mechanisation was continuing, with some mines introducing mechanical coal winning and haulage. However the Board went on to report that of the 54 underground mines, only 14 were fully mechanised, 6 partly mechanised, while the remaining 34 were "entirely non-mechanised."[645] Murray, in his history of the West Moreton coalfield, claimed that it was only in 1961-62 that the QCB and the Mines Department realised that many mines were still being worked by hand.[646] However, as Murray also noted, the QCB's policy of letting the small inefficient underground mines simply fade away was having an effect.[647] In 1953 there were 89 underground mines in the State; the number was down to 65 by 1961. By 1970 there were only 33 underground mines operating, and many of these would close during the 1970s or 1980s. Employment in the underground mines in Queensland peaked in 1955 at just over 3500, but by 1960 had fallen to 3074; employment reached a low of just under 1500 in 1969 before it began to rise once again.

Concerns emerge over export prices to Japan

By the late 1960s, concerns in Australia over the prices we received for our coal exported to Japan began to be expressed, but the growth in the Japanese steel industry, the prospect of rapidly increasing export tonnages and concerns over the capacity of the rail and port infrastructure in NSW to handle significantly higher exports tended to receive more public and political attention. In November 1968 in its annual report for the 1967-68 year, the JCB said that Australian coal was being sold too cheaply and that substantial price increases should be required in future export contracts to Japan. "The truth is that NSW coal, being so cheap, the Japanese steel industry with its expanding requirements has been willing to take all the coal that has been available."[648] The JCB quoted average prices of coal imported by Japan in July from its six major suppliers. The average price for coal from the USA was US$19.57, from Canada US$15.38 and from Australia US$12.80. The price received by Australia was also below the level for Russia and Poland, and only exceeded the level for China. The JCB said that "For some time now, the board has had doubts as to whether, so afar as the Australian coal mining industry is concerned, the national interest is best served by leaving the sale of coal to the ingenuity of exporters competing in the market of one country… There would be even more doubt in the event that the market were to show any contraction."

The message from the JCB to the Federal and NSW Governments was clear: don't trust the coal exporters to get a reasonable price, and government intervention in the coal negotiations was necessary. The NSW coal companies reacted strongly to the JCB's statements, with their chairman, Sir Edward Warren, slamming the JCB as not qualified to criticise the efforts of those who had worked for years to develop a coal export trade. Warren said that the coal companies were angry at suggestions that exporters had let down the nation and presumably their own shareholders. He said that prices depended on market conditions at the time and that coal negotiators were convinced that they consistently obtained the best possible prices given the conditions in the market.[649] The JCB's concern over export prices featured again in its annual report for 1968-69. In this report it noted the sophisticated approach to coal contract negotiations by the Japanese, and the strong cohesion between the various companies in Japan. The JCB said that Australia needed to match the Japanese approach as there was no comparable sophistication and cohesion by Australia in negotiating contracts, with export prices suffering as a consequence.[650] The JCB had made a detailed evaluation of each brand of coal exported from NSW to Japan and had advised each coal company what price it believed it could obtain. Subsequent negotiations had resulted in increases of up to $1 per ton, but the JCB said it had expected more.

In March 1969, during negotiations in Tokyo for the new Japanese financial year commencing the following month, Clutha Development's senior executive, George Jennings, actually threatened to cease supplying coal to the JSM from 1 April unless the JSM agreed to an increase of $1.50 per ton. The JSM had already settled with other exporters including RW Miller and Gollin for a $0.60 per ton increase and rejected the Clutha demand. Jennings then broke off the negotiations and returned to

Australia, although the company later withdrew its demand and resumed negotiations.[651] The JCB came out publicly in support of Clutha, with its Chairman, Bernard Hartnell, saying that having regard to the prices the JSM were prepared to pay exporters in other coal exporting countries, Clutha was certainly entitled to an increase of $1.50 or more. Hartnell said that the national interest was not well served if prices remained at unduly low levels.[652]

But the growth in the Japanese steel industry, and the potential for Australia to become an even more important supplier, continued to be prominent in the media. In December 1969, NSW Mines Minister Wal Fife visited Tokyo for talks with the JSM and was told that they would like to see coal exports from NSW double by 1975. The JSM also referred to plans in NSW to increase coal loading facilities to cater for such an increase, encouraging the NSW Government to plan for even greater capacity with the likelihood that Japanese demand would be greater than expected.[653] The question of the fairness export prices Australia was receiving for its coal would become a major issue in Australia in the 1970s as demand from Japan continued to surge, although a recession in the Japanese steel industry in 1974 and 1975 would cause shockwaves in Australia as contract tonnages previously thought secure would prove to be "flexible".

Coal regions – some grow, some struggle

The changing market conditions in the 1960s had major impacts on the different coal districts in NSW and Queensland, and these impacts were particularly evident from the employment trends in each of the major areas. In Queensland employment dropped from 3245 in 1960 to only 2294 in 1970, largely due to the dramatic decline in the West Moreton where job numbers fell from 1998 to 894.[654] There were major falls also in the Darling Downs district (143 to 11), Rockhampton (184 to zero), and Bowen (541 to 196). The Bowen Basin however had begun its growth period with employment rising from only 102 in 1960 to 1029 by 1970; in the Bowen Basin, the developments at Blackwater and Kianga/Moura were generating all the new jobs. The period from 1968 to 1970 was the low point for overall employment in Queensland and the 1970s would see job numbers rising rapidly as the Utah and other projects in the Bowen Basin commenced.

NSW also saw very different trends in each district, with the South Maitland and Western districts experiencing significant reductions in employment in coal mining. Between 1960 and 1970, the South Maitland district (the district where mining from the Greta seams dominated) employment dropped from 3245 to only 1435. In the Western district the reduction was also dramatic, from 1050 to 585. In contrast the other districts saw significant increases, with Newcastle up from 3813 to 4641, the North West (the upper Hunter Valley and Gunnedah areas) up from 761 to 1036, the South Coast up from 3853 to 4955, and the Burragorang Valley up from 584 to 1157. However overall employment in NSW rose however by less than 500 over the ten years.

Production and employment in the South Maitland district was supported in the 1960s by the export market. Production by 1969-70 was around 2.2 million tons and had been relatively steady since the mid-1960s. Sales for town gas in NSW had fallen to only 300,000 tons by 1970, and sales to other states were down to around 150,000 tons. Exports however were on the rise, reaching 1.6 million tons in 1970, with Japan taking the bulk of the exported tons. In Japan the South Maitland coal was used as a high volatile soft coking coal by the gas, chemical and coke industries. By 1970 the district's mines, mostly working the Greta seams, were concentrated around Cessnock and in the area south of Cessnock, with one mine still operating near Branxton.

The next 50 years

The 1960s can be seen as the decade during which the Australian coal industry began its reorientation to the export market; it was arguably the time when the industry in NSW and Queensland was born as a major modern export industry. The decades ahead would prove to be tough, but the industry would emerge as the world's leader in terms of export tonnages and a leader in mine safety.

The second volume of *Coal: the Australian Story* will continue the story of the Australian coal industry from the 1970s to the present day. The 1970s, 1980s and 1990s were turbulent periods, with the 1970s featuring the oil shocks of 1973 and 1979, the major currency realignments which impacted the US dollar and the yen, the cessation of growth in the Japanese steel industry, the controversies about coal prices, and the policies of the Whitlam Labor Government including export controls, controls on foreign investment and the coal export duty. However the decade also saw the emergence of a new international trade in steaming coal which held bright prospects for Australia. By 1980, coal was seen as having to play a critical role in the generation of electricity throughout the world, and in particular in Japan, in other developing Asian economies and in Europe. Oil had become very expensive and many countries were fearful of their reliance on the unstable Middle East for crucial oil supplies.

But by the early 1980s, the expected boom in the demand for commodities, and in particular coal, began to be seen as something of a mirage. Over optimistic forecasts of demand were pared back, but the decade was to see over- production, jobs losses, fierce competition, low prices and poor or negative profits. The time had also come when the producers began to pressure the Federal and State Governments for major reforms in the way in which the coal industry was regulated and taxed. Changes were made to the industrial relations system in Australia, but not to the coal industry, which remained cocooned in its own system, quite separate from other industries. The demise of the Coal Industry Tribunal, established in 1947, would come in 1995, when the Keating Labor Government would absorb the Tribunal into the Industrial Relations Commission. The 1990s would also see a major reduction in the powers of the Joint Coal Board, and by the end of the decade, moves which would later culminate in the transfer its remaining functions to a company owned by the industry itself. The Queensland Coal Board would also be abolished, ending its near 50 year existence. While the CIT would be abolished under a Labor Government, the election of the Howard Coalition Government in 1996 would see a fundamental overhaul of the mainstream industrial relations system, facilitating major restructuring of the coal industry. The 1990s would see some of the longest lasting and most bitter industrial disputes in the industry's history involving major companies such as Rio Tinto and Arco. Major restructuring of operations at many minesites would see significant cutbacks in employment, but also major improvements in productivity in the second half of the 1990s. After tragedies involving multiple fatalities at mines including Gretley near Newcastle and Moura in the Bowen Basin, mine safety legislation and regulation in both NSW and Queensland would be completely overhauled and modernised.

The years since 2000 have also been eventful, with the growth of China beginning to impact on commodity markets in a major way from around 2004. China, then a significant coal exporter, began to look to imported coal to fuel its steel mills and power stations, and by 2009, coal prices had hit record highs. The emergence of China and India as major importers and the growth in other markets have seen Japan's role become far less dominant. Japan is still a major market for Australia's metallurgical and thermal coal, but the controversies of the 1970s, 1980s and 1990s over coal prices have now tended to fade away. The controversies in the new era are now more often related to climate change and the future role of coal. Major companies have exited coal mining, including Rio Tinto. BHP has reduced its thermal coal mine stable in Australia to the Mt Arthur operation in the Hunter Valley. South 32 is planning to move out of thermal coal mining in South Africa, and Glencore has imposed a limit on its global production following pressure from investors and climate change activists. The future is becoming much more uncertain for coal mining in Australia.

Abbreviations – Endnotes

ACA – Australian Coal Association

AFR – Australian Financial Review

BC – Brisbane Courier

CT – Canberra Times

CM – Courier Mail

CPD – Commonwealth Parliamentary Debates

DT – Daily Telegraph

DMR – NSW Department of Mineral Resources

IEA – International Energy Agency

HRA – Historical Records of Australia

HRNSW – Historical Records of New South Wales

IM – Illawarra Mercury

JCB – Joint Coal Board

MM – Maitland Mercury

NAA – National Archives of Australia

NMH – Newcastle Morning Herald; also Newcastle Morning Herald and Miners' Advocate

NSW – New South Wales

NSWCA – New South Wales Coal Association

NSWCCPA – NSW Combined Colliery Proprietors' Association

NSWMC – NSW Minerals Council

NSWPD – NSW Parliamentary Debates

OECD – Organisation for Economic Cooperation and Development

QCB – Queensland Coal Board

QPD – Queensland Parliamentary Debates

QT – Queensland Times (Ipswich)

QGMJ – Queensland Government Mining Journal

QMC – Queensland Mining Council

QRC – Queensland Resources Council

REQ – Resources and Energy Quarterly, Department of Industry, Innovation and Science

SMH – Sydney Morning Herald

WA – Western Australia

Endnotes

Introduction and Chapter 1

1   https://hunterlivinghistories.com/history/; accessed 2 January 2018

2   ibid

3   M H Ellis, A Saga of Coal, p.4

4   A Cousins, The Garden of New South Wales, p.14

5   HRA Series 1 Volume 11 p.33, Hunter to Duke of Portland 25 June 1797

6   HRNSW Vol.3 p.289, Bass to Paterson 20 August 1979

7   HRA Series 1 Volume 11 p.118, Hunter to Duke of Portland 10 January 1798

8   HRNSW Volume 3 pp.716-717 8 Sept 1799

9   Turner, Coal Mining In Newcastle 1801 to 1900, p.14

10   HRNSW Volume 4 p.359 King to Banks April 1801

11   HRA Series 1 Volume 111 p.116 King to Portland 8July 1801

12   HRA Series 1 Volume 111 p.14 King to Portland 10 March 1801

13   Ellis, p.15

14   Turner, Coal Mining, p.16

15   Ellis, p.19

16   E Ross, History of the Miners' Federation of Australia, quoted in R Taylor, In The Black – The Auscoal Super Story, p.5

17   Turner, Coal Mining, p.19

18   See J W Turner (ed); Newcastle as a Convict Settlement: The Evidence before J T Bigge, p.98

19   Turner Evidence, pp138-142

20   Bigge Report, pp.114-118

21   The Australian, 1 April 1826, p.2

22   Ellis, p.25 – Sir George Murray, Colonial Office Secretary to Governor Darling, 31 July 1828

23   Gregson, *The Australian Agricultural Company* , p.51

24   Gregson, p.60

25   Gregson, p.62

26   Turner , Coal Mining ,p.40

27   Turner , Coal Mining, p.41

28   Turner, Coal Mining ,p.41

29   Turner, Coal Mining, p.43

30   C Jay. *The Coal Masters*, p.16

31   Jay, p.21

32   Jay, pp.32-33

33   NMH 18 August 1877, p.4

34   J Gunn, Along Parallel Lines: A History of New South Wales Railways, p.65

35   Gregson, pp.154-155

36   Turner, Coal Mining ,p.59

37   Gregson, p.154

38   Sydney Morning Herald 10 September 1849 p.6

39   A Cremin, The Growth of an Industrial Valley: Lithgow, New South Wales, p.36

40   Cremin, p.36

41   Cremin, pp.36,37

42   E F Dunne, Brief History of the Coal Mining Industry in Queensland, p.320

43   Dunne, p.321

44   Dunne, p.327

45   Brisbane Courier, 1 September 1885, p.3

46   Brisbane Courier, 15 December 1891, p.6

47   Australian Dictionary of Biography

48   Murray, p.6

49   See http://www.parliament.uk/about/living-heritage/transformingsociety/livinglearning/19thcentury/overview/coalmines/ accessed 26Nov2012

50   Gollan,The Coalminers of New South Wales, pp.17-18

51   Maitland Mercury 26 May 1857, p.3

52   ibid

53 Ross E, A History of the Miners' Federation, p.18

54 Gollan, p.36

55 Newcastle Chronicle and HRD News, 29 May 1861, p.2

56 Gollan p.37

57 Maitland Mercury 22 August 1861

58 SMH 2 September 1961, p.5

59 Ellis, p.83

60 Gollan. P.42

61 Gregson, p.239

62 Gollan, p.44

63 Turner p.71

64 Fleming (in D. Merrett (ed) Business Institutions and Behaviour in Australia, p.52) states that the first agreement to fix prices was between AACo and Newcastle Coal & Copper Co in 1855

65 Gollan, p.46

66 Jay, p.28

67 Gregson, p.254

68 Turner, Coal Mining, p.76

69 Fleming, p.54

70 Turner, Coal Mining ,p.79

71 Turner, Coal Mining ,p.80

72 Turner, Coal Mining, p.80

73 Ellis, p.106

74 Fleming, p.55

75 SMH 24 March 1874, p.3; Maitland Mercury 24 March 1874, p.2

76 SMH 24 March 1874, p.3

77 Maitland Mercury, 24 March 1874, p.2

78 Fleming, pp.59-60

79 Ellis, p.105

80 Ellis, p.105

81 Gregson, p.273. This comment was from a mining engineer from England who was appointed by the company as a consulting colliery engineer to review operations in 1874

82   SMH 2 June 1880, p.7

83   SMH 3 October 1884

84   Christison p.23

85   Turner, p.100

86   Gollan, p.67

87   Gollan p.68

88   Reagan and Kininmonth, p.520

89   R L Whitmore, Coal in Queensland – the first fifty years, pp.165-166

90   NSW Department of Mines Annual Report, 1881, p.115

91   D Dingsdag, The Bulli Mining Disaster (1993) p.39; see also SMH 31
      October 1902 p.8

92   Illawarra Mercury, 3 May 1887. P.2

93   Dingsdag (1993)

94   Dingsdag (1993) p.11

95   Dingsdag (1988) p. 41

96   SMH 13 July 1887 p.13

97   SMH, 30 August 1888, p.5

98   Gollan, Coalminers of NSW, p.85

99   Gollan, Coalminers of NSW, p.86

100   Gollan, p.107

101   See Two Depressions, One Banking Collapse, RBA Research
      Discussion Paper 1999-06

102   Gollan, pp.89-90

103   Gollan, p.88

Chapter 2

104   C H Martin et al, History of Coal Mining in Australia, pp.36,42

105   Metropolitan is still an operating mine, part of the Peabody Energy
      group.

106   F. Brady, Electricity in NSW, p.7

107   K Burley, The Economic Record, Vol 36, August 1960, p.394

108   Burley p.394

109   From J W Turner in Australian Dictionary of Biography, Volume 7,
      (MUP), 1979

110 H L Wilkinson, The Trust Movement in Australia, pp.52-53

111 These figures are from Ellis, A Saga of Coal, p.175

112 NMH 2 May 1900, p.3

113 Brisbane Courier 17 August 1900, p.5

114 Brisbane Courier, 4 July 1900, p.6

115 Brisbane Courier, 3 July 1900, p.4

116 Brisbane Courier, 4 June 1901, p.4

117 Brisbane Courier, 19 September 1908, p.5

118 Queensland Times, Ipswich Herald and General Advertiser, 16 March 1907, p.5

119 R Whitmore, Coal in Queensland: From Federation to the Twenties, p.7

120 The Queensland Official Year Book 1901, p.125

121 Brisbane Courier, 7 February 1902, p.48

122 Based on data in Queensland Department of Mines Annual Reports

123 These death rates are from the NSW Department of Mines Annual Report 1900, p.131

124 Ibid, page following p.92

125 Queensland Department of Mines Annual Report, 1901, p.126

126 NSW Mines Department Annual Report 1900, pp.116,123

127 For example, see NSW Department of Mines Annual Report 1900, p.170

128 N R Monger, The Newcastle Coalfield – 100 years of change, p.388

129 The Royal Commission into the disaster listed 95 men and boys killed. Piggin & Lee in *The Mount Kembla Disaster* list 96, with one miner later succumbing to his injuries.

130 Piggin & Lee, The Mt Kembla Disaster, p.46

131 NMH, 30 October 1897, p.5. The article says that 82 applicants were from England, but it is likely that some of these were from Scotland and Wales.

132 NSW Department of Mines Annual Report, 1897, p.118

133 Piggin & Lee p.199

134 Piggin & Lee p.199

135 Piggin & Lee p.158

136   Report of the Royal Commission on the Mount Kembla Colliery Disaster 1903

137   NSW Department of Mines Annual Report 1902 p.60

138   Piggin & Lee p.198

139   Piggin & Lee p.214

140   Piggin & lee p.218

141   NSW Department of Mines Annual Report 1903 p.100

142   NSW Department of Mines Annual Report 1903 p.100

143   Ibid p.100

144   NSW Department of Mines Annual Report 1905 p.106

145   NSW Department of Mines Annual Report 1909 p.131

146   NSW Department of Mines Annual Report 1912 p.119

147   NSW Department of Mines Annual Report 1905 p.105

148   NSW Department of Mines Annual Report 1888, p.32

149   From 1902 to 1911 Pelaw Main was registered to John and William Brown, not the firm J&A Brown. It was transferred to J&A Brown in 1912. See C Jay *The Coal Masters* p.44

150   NSW Department of Mines annual reports

151   NMH 18 November 1904, p.5

152   http://hosting.collectionsaustralia.net/newcastle/greta/pelaw.html

153   Jay, p.70

154   NSW Department of Mines Annual Report 1913, p.163

155   Jay, p.52

156   ibid

157   QT, 11 March 1905,p.13

158   ibid

159   Queensland Department of Mines Annual Report 1911, p.169

160   Gollin, p.106

161   Gollin, pp.110-111 (based on AACo meeting minutes)

162   Gollin, p.127

163   B Bowden, Work and Strife in Paradise, p.15

164   Gollin, p.126

165   G Freudenberg, Cause for Power, p.184

166   QT, 11 April 1878, p.3

167   QT 16 October 1883, p.4 mentions half yearly meetings of the Queensland Coal Miners' Association.

168   QT, 30 January 1891, p.3

169   QT, 9 June 1900, p.5

170   QT, 10 April 1902, p.7

171   Dunne, Brief History of Coal Mining in Queensland, p.337

172   Brisbane Courier, 17 July 1908, p.3

173   J Bach, *A Maritime History of Australia,* p.211

174   Bach pp.211-212; based on Huddart Parker company papers

175   Gollan *The Coalminers of New South Wales p.119*

176   D Dingsdag, The Restructuring of the NSW Coal Industry 1903-1982, p.62

177   Bach p.212

178   Bach p.212

179   H L Wilkinson, *The Trust Movement in Australia* p.79

180   NMH 11 October 1906, p.8

181   CPD, 3July1906

182   SMH 1October 1907 p.6

183   Ellis pp.160-161

184   R v Associated Northern Collieries [1911] HCA 73; (1911) 14 CLR 387 (22 December 1911)

185   Adelaide Steamship Co Ltd v R [1912] HCA 58; (1912) 15 CLR 65 (20 September 1912)

186   Attorney-General (Cth) v Adelaide Steamship Company Limited [1913] UKPCHCA 2; (1913) 18 CLR 30 (25 July 1913)

187   Report of the (Campbell) Royal Commission to Inquire into the Coal Industry 1919, p.43

188   SMH, 22 March 1915, p.8

189   NMH 1 March 1911, p.6

190   NSW Department of Mines annual reports

191   D J Murphy chapter, Thomas Joseph Ryan, in Murphy (ed) *The Premiers of Queensland,* p.278

192   The SMH reported that the Government was negotiating to buy Box
       Flat, Parkhead, Aberdare, and Abermain Extended mines on 25 January
       1916.

193   In 1907 the Government commissioned the sinking of a shaft on
       Crown land near the Dawson River to extract 200 tons of coal for
       testing by the Royal Navy. While not an operating coal mine, it was
       perhaps the first Government coal operation in Queensland. Refer
       Whitmore (Queensland Coal: From Federation to the twenties) p.22

194   NMH 21 February 1913, p.5

195   NMH 23 March 1915, p.5

196   Jay, p.111

197   ibid

198   Gollan, Coalminers, p.147

199   Gollan, Coalminers, p.147

200   Davidson, Coal in Australia, pp.548-549

201   Gollin, p.152

202   SMH 20 September 1917, p.3

203   Gollin, p.156

Chapter 3

204   Adelaide Advertiser, 14 April 1919, p.7

205   Ibid

206   Second and Final Report of the Royal Commission of Inquiry into the
       Coal Mining Industry and the Coal Trade in the State of New South
       Wales, 1920, p6. (Campbell Report 1920)

207   Campbell Report 1920 p.9

208   Report of the Royal Commission of Inquiry into the Coal Mining
       Industry and the Coal Trade in the State of New South Wales, 1919
       p.43 (Campbell Report 1919)

209   Campbell Report 1919 p.45

210   Campbell Report 1919 p.49

211   SMH 20 August 1914 p.9

212   Campbell Report 1919 p. 14

213   Gollan, p.169

214   SMH 24 March 1923, p.15

215 QGMJ no. 48160 December 1923, p.450

216 NMH 14 July 1926, p.9

217 Whitmore, Federation to the Twenties, p.242

218 ibid

219 Whitmore, Federation to the Twenties, p.243

220 Whitmore, Federation to the Twenties, p.439

221 Queensland Department of Mines Annual Report 1920, p.121

222 Queensland Department of Mines Annual Report 1933, p.2

223 J Lang, The Great Bust p.114

224 Lang, p.119

225 SMH 12 September 1927, p.15

226 Gollin, pp.186-187

227 ibid

228 Gollin. P.188

229 See SMH 12 April 1929, p.13; also 15 August, p.11; 23 July, p.11; 18 May, p.17

230 A Metcalfe, For Freedom and Dignity, p.147

231 NSW Department of Mines Annual Report 1938, p.15

232 SMH 19 June 1929, p.16

233 SMH 20 June 1929, p.11

234 Lang, pp.115-116

235 F R Mauldon, The Economics of Australian Coal, p.79

236 Davidson RC, p.320

237 Davidson RC, p.110

238 Davidson RC, pp.258-259

239 Davidson RC, pp.314-315

240 Gollin, pp.178-179

241 Davidson RC, p.352

242 Davidson RC, p.325

243 SMH 28 November 1930, p.12

244 Year Book Australia 1935, p.373. These figures are for members of trade unions.

245 Ellis, pp.205-206

246   NMH, 31 May 1930, p.15

247   QGMJ, September 1929, p.408

248   Davidson Inquiry 1946, Vol.2, p.374

249   ibid

250   Davidson Inquiry 1946, Vol.2, p.375

251   Davidson Inquiry 1946, Vol.2, p.376

252   Courier Mail 11 November 1933, p.15

253   ibid

254   These are average "pit mouth" values from Mines Department annual reports.

255   Courier Mail 26, 27, 28 and 29 May 1936.

256   QT 8 August 1936, p.10

257   CM, 26 May 1936, p.12

258   Telegraph (Brisbane), 4 June 1936, p.7

259   CM, 29 May 1936, p.14

260   H Hughes, The Australian Iron and Steel Industry, p.99

261   The NSW Mines Department annual report for 1936 reported (p.76) that 3 Jeffrey electrically operated loaders had been installed in Lambton. It also reported that the Wallarah colliery had ordered 6 Joy electric loaders, 3 of which had been received, but which had not yet been put into use.

262   BHP, Steel Industry Collieries Development, p.3

263   NSW Department of Mines Annual Report 1936, p.76

264   ibid

265   BHP, Steel Industry Collieries Development, p.3

266   E Ross, A History of the Miners' Federation of Australia, p.368

267   E Ross, p.354

268   Orr and Nelson. Coal: The Struggle of the Mineworkers, p.2

269   Orr and Nelson, p.3

270   Orr and Nelson, p.3

271   Orr and Nelson, p.12

272   Gollan, pp203-204; Orr and Nelson, p.12

273   Orr and Nelson, p.27

274 NSW Year Books, 1934-35 and 1945-46

275 The average shifts per week are simply based on dividing the annual average by 5.2, assuming 260 days per year as the maximum number

276 QGMJ July 1933, p.205

277 QGMJ July 1935, p.243

278 Brisbane Courier, 24 February 1931, p.9

279 Courier Mail, 21 September 1935, p.11

280 Gollan, pp.212-213

281 NMH 19 December 1939, p.7

282 A Murray, No Easy Field, p.58; QT 2 June 1928, p.10

283 Queensland Department of Mines Annual Report 1911, p.70

284 NSW Department of Mines Annual Report 1937, p.84

285 NSW Department of Mines Annual Report 1937, p.92

286 NSW Department of Mines Annual Report 1938, p.77

287 NSW Department of Mines Annual Report 1937, p.86

288 NSW Department of Mines Annual Reports 1927-28 and 1937-38

289 Ellis, p.207

290 Commonwealth Court of Conciliation and Arbitration, Print No. 3811 pp.7-8

291 Ellis, p.214

292 NMH 1 January 1939, p.7

Chapter 4

293 Gollin, p.218

294 For a history of the miners' pension schemes see Ross Taylor's *In the Black: The Auscoal Super Story.*

295 NMH, 12 December 1941, p.18

296 NMH, 29 December 1941, p.2

297 Telegraph (Brisbane) 9 July 1942, p.6

298 Taylor, p.19

299 Taylor, p.17

300 SMH, 16 May 1940 p.9

301 This section is based on the SMH article 20 April 1940 p.17

302   Courier Mail, 8 March 1940, p.1; Murray, p.135

303   NSW Year Book 1939-40, p.629

304   See S J Butlin, War Economy 1939-1942, p. 410

305   Butlin, War Economy, p.410

306   Butlin, War Economy, p.416

307   SMH 8 August 1941, p.5

308   This court was the forerunner to the Commonwealth Conciliation and Arbitration Commission and the Australian Industrial Relations Commission.

309   Davidson table opposite p. 32

310   S J Butlin and C V Schedvin, War Economy 1942-1945, p.443

311   Telegraph (Brisbane) 9 February 1942, p.3

312   Butlin and Schedvin, p. 444

313   Butlin and Schedvin. p.444

314   CPD, 8 May 1942

315   CPD, 13 May 1942

316   Butlin and Schedvin, p. 445

317   Butlin and Schedvin, p.447

318   SMH 12 October 1943, p.7

319   With the independence of the local boards, Butlin (p.450) concluded that "Plainly the intention of the Miners' Federation was to play the decision of one local board against the others…"

320   West Australian, 3 May 1943, p.2

321   CPD 8 March 1944, Vol. 177 p.1065

322   Butlin and Schedvin, p.452

Chapter 5

323   Davidson 1946, table opposite p.32

324   NSW Department of Mines Annual Report 1946, p.11

325   Queensland Department of Mines Annual Report 1946, p.83; Powell Duffryn, Vol.1, pp.32,52

326   Martin et al, p.96

327   Davidson 1946, pp243-45

328   Davidson 1946, p.390

329 JCB Annual Report 1947-48, p.17

330 JCB Annual Report 1966-67, p.181

331 Davidson 1946, p.327

332 Davidson 1946 p.238

333 Martin et al, pp.122-123

334 Powell Duffryn Technical Services, First Report on the Coal Industry of Queensland, 1949, p.123

335 H Elford and N McKeown, Coal Mining in Australia.

336 Elford and McKeown p.35

337 Elford and McKeown, pp.133-137

338 Elford and McKeown, pp.137-142

339 Elford and McKeown, pp. 162-164

340 The more common method was to last the coal from the face once the seam had been under-cut.

341 Elford and McKeown, pp.180-182

342 Elford and McKeown, pp.184-188

343 Jigs were devices which used the power of full skips descending a slope to haul empty skips up the slope.

344 A jig was also a term for a device which removed stone from the coal.

345 Elford and McKeown, pp.184-188

346 Elford and McKeown, p. 240. This section relates to mainly to NSW. Queensland conditions may have differed slightly.

347 Powell Duffryn, p.123

Chapter 6

348 Davidson 1946, p. 12

349 Davidson 1946, p. 12

350 Davidson 1946, p. 12

351 Davidson 1946, p.339

352 Davidson 1946, p.48

353 Davidson, p.248

354 NMH, 3 May 1946, p.4

355 Ross, p.399

356 CT, 29 August 1945, p.2

357   NMH, 3 May 1946, p.1

358   SMH 28 May 1946, p.1

359   SMH 29 May 1946, p.1

360   SMH 5 June 1946, p.1

361   C Fisher, Coal and the State, p.108

362   CPD, 30 July 1946

363   CPD, 24 July 1946

364   NSWPD 28 August 1946, p.40

365   NSWPD 28 August 1946, p.42

366   NSWPD 28 August 1946, p.45

367   NSWPD 28 August 1946, pp.51-53

368   NMH 19 September 1946, p.3

369   NMH 30 July 1946, p.3

370   ibid

371   CPD, 30 July 1946

372   Jay, p.155

373   Common Cause, 11 January 1947.

374   JCB Annual Report 1947-48 pp.6-7

375   ibid

376   JCB Annual Report 1947-48 p.5

377   JCB Annual Report 1947-48, pp.5-6

378   JCB Annual Report 1947-48, pp.6-7

379   JCB Orders were legal instruments under the Coal Industry
      Acts, equivalent to a regulation.

380   JCB Annual Report 1947-48 p.12

381   JCB Annual Report 1948-49 p.7

382   C Jay The Coal Masters p.157

383   SMH 24 August 1948 p.3

384   JCB Annual Report 1949-50 p.8

385   JCB Annual Report 1948-48 p.12

386   JCB Annual Report 1949-50 p.13

387   ibid

388   Courier Mail 22 November 1947 p.3

389 Courier Mail 30 January 1948 p.3

390 Powell Duffryn, Vol.1 p.26

391 Powell Duffryn, Vol.1, p.141

392 Powell Duffryn, Vol.1, p.27

393 Powell Duffryn, Vol.1, p.28

394 Powell Duffryn, Vol.1, pp.61, 63-64

395 Powell Duffryn, Vol.1, pp.71,73,85,92

396 R J Kemp, Report on the Large Scale Development of the Blair Athol Coalfield, p.8

397 Courier Mail, 4 December 1947, p.3

398 Courier Mail, 29 January 1948, p.3

399 Dunne , pp.325-326

400 SMH 20 February 1942, p.4

Chapter 7

401 Based on data from NSW Year Books for average employment and total wages and salaries.

402 NMH, 14 October 1946, p.3

403 NMH 14 April 1948, p.3

404 Queensland Times, 20 April 1948, p.2

405 For example in 1946 Coal mining accounted for approximately two thirds of NSW employment in mining and quarrying NSW Year Book pp. 676, 751

406 SMH 23 June 1948, p.1.

407 NMH 29 June 1948, p.3

408 SMH 6 July 1948. P.1

409 For example the SMH on 6July1948 p.2 referred to the Federation's leadership and the "Communist sabotage of the coal mining industry"

410 http://www.historyguide.org/europe/churchill.html

411 F Daly, From Curtain to Kerr, p.67

412 Day pp. 477-488

413 Day p.488

414 Daly p. 67

415 Ross, p.104

416   Day, p.491

417   Courier Mail, 12 July 1949, p.1

418   JCB Annual Report 1948-49, p.33

419   JCB Annual Report 1948-49. P.34

420   JCB Annual Report 1948-49, p.33

421   Ross, p.102

422   Metcalfe, p.156

423   NSW Year Book 1945-46, p.787

424   JCB Annual Report 1971-72, p.72

425   QCB annual reports

426   SMH 19 September 1947, p.2

427   Sun, 27 August 1944, p.3

428   Coaltown: A Social Survey of Cessnock

429   Walker, pp.4, 126-127

430   Walker, p.38

431   Walker, p.39

432   Metcalfe, pp.79-80

433   Metcalfe, p.80

434   Jay, p.150

435   Quoted in Davidson, Coal in Australia, 1953, p.557

436   SMH 7 July 1948, p.2

437   JCB Annual Report 1947-48, p.25

438   JCB Annual Report 1949-50, p.33

439   JCB Annual Report 1947-48, pp.6-7

Chapter 8

440   JCB Annual Report 1950-51, p.15

441   H Hughes, The Australian Iron and Steel Industry 1848-1962, pp.145-146

442   ibid

443   JCB Annual Report 1949-50, pp.18-19

444   NMH 8 June 1950, p.5

445   NMH, 26 August 1953, p.2

446   NMH 17 March 1954, p.8
447   NSW Department of Mines Annual Report 1959, p.55
448   SMH 7 January 1953, p.2
449   The Coal Miner, June 1952, pp.16-17
450   QCB First Annual Report 1949 to 1952, p.10
451   Ibid, p.11
452   W L Hawthorne, Coal Exploration by the Queensland Department of
      Mines, p.16
453   QCB, First Annual Report 1949 to 1952, pp.22-23
454   QCB Annual Report 1952, p.12
455   QCB Annual Report 1954-55, p.20
456   Brisbane Telegraph, 8 February 1951, p.9
457   CT, 10 July 1952, p.4. Murray, p211
458   Article by E Warren, SMH 7 January 1952, p.2
459   SMH 8 June 1948 p. 1
460   SMH 21 August 1948 p. 3
461   ibid
462   JCB Annual Report 1950-51, p.21
463   ibid
464   JCB Annual Report 1949-50, p.34
465   JCB Annual Report 1949-50, p.33
466   NMH 25 November 1952,p.2
467   NMH 20 November 1952, p.3
468   NMH 26 August 1954, p.1
469   N R Monger, The Newcastle Coalfield – 100 years of change, in
      AusIMM Centenary Conference Proceedings 1991, p.391
470   JCB annual reports
471   JCB Annual Report 1947-48, p.23
472   NMH 13 November 1952, p.2
473   JCB Annual Report 1959-60, p.73
474   NMH 21 December 1950, p.1
475   SMH 22 December 1950, p.3
476   NMH 22 January 1951, p.3

477 NMH 25 January 1951, p.2

478 IM 14 March 1951, p.2

479 SMH 3 March 1951, p.1

480 SMH 13 March 1951, p.4

481 SMH 12 May 1951, p.3

482 NMH 24 May 1951, p.3

483 Jay, p.166

484 Dingsdag, p.406

485 Dingsdag, pp.409-410

486 JCB Annual Report 1957-58, p.12

487 SMH 30 July 1960, p.8

488 JCB Annual Report 1959-60, p.27

489 QCB Annual Report 1955-56, p.23

490 QCB Annual Report 1959-60, p.17

491 QCB Annual Report 1959-60, pp.6,26

492 First Report of the Queensland Coal Board, 1949 to 1952, p.15

493 QCB Annual Report 1956-57, p.2

494 NMH 9 July 1952, p.3

495 JCB Annual Report 1948-49, p.6

496 JCB Annual Report 1949-50, p.8

497 JCB Review of Activities, p.140

498 JCB Annual Report 1949-50, pp.10-11

499 NMH 12 November 1952, p.2

500 ibid

501 JCB Annual Report 1952-53, p.34

502 JCB Annual Report 1947-48, p.27

503 JCB Annual Report 1952-53, p.18

504 JCB Annual Report 1952-53, p.17

505 Howard J, The Menzies Era, p.230

506 Watt A, The Evolution of Australian Foreign Policy 1938-1965, p.217

507 Howard, p.234

508 Howard, p.239

509 NSW Department of Mines Annual Report 1904, p.112

510 T Okazaki, Productivity Change and Mine Dynamics: The Coal Industry in Japan during World War II, p.31

511 I Kume, Disparaged Success: Labour Politics in Post War Japan, p.155

512 ibid

513 J Price, Japan Works: Power and Paradox in Postwar Industrial Relations p.233

514 M Sumiya (ed), A History of Japanese Trade and Industry Policy, p.313

515 M Sumiya (ed), A History of Japanese Trade and Industry Policy, p.314

516 Kume, p.159

517 The Miike mine also saw Japan's worst mining tragedy in 1963 when a coal dust explosion killed 458 miners.

518 Price, p.235

519 Price, p. 236.

520 Price, p.236

521 Sydney Morning Herald, 3 November 1953, p.6

522 Sydney Morning Herald, 27 December 1954, p.2

523 QGMJ February 1968

524 D Lee, The Second Rush, p.44

525 The details here are from Jay, pp.175-176

526 Jay, p.175

527 Jay, pp.174-176

528 SMH, 19 February 1959, p.3

529 ibid

530 JCB Annual Report 1955-56, p.7

531 CT 1 March 1956, p.2

532 CT 20 March 1956, p.1

533 SMH 6 November 1956, p.14

534 SMH 27 July 1957, p.1

535 SMH 20 July 1957, p.1

536 SMH, 9 February 1957, p.4

537 SMH 16 September 1958, p.4

538 R Broomham, First Light: 150 Years of Gas, p.178

539 Broomham, p.159

540   CT 15 February 1958, p.3

541   QGMJ Vol.78 February 1958, p.87

542   JCB Review of Activities 1962, section 104 - quoted Annual Report 1950-51

543   JCB Review of Activities 1962, section 102

544   ibid

545   QCB Annual Report 1953-54, p.21

546   ibid

547   QCB Annual Report 1953-54, p.10

548   JCB Annual Report 1958-59, p.44

549   NMH 21 July 1950, p.4

550   Barrier Daily Truth, 24 October 1947, p.1

551   www.coalservices.com.au accessed 171117

552   NSW Department of Mines Annual Reports for 1951, p.54 and 1952, p.57

553   NSW Department of Mines Annual Report 1951, p.54

554   JCB Annual Report 1966-67, p.181; recurrent bronchitis included chronic obstructive lung disease.

555   JCB Annual Report 1947-48, p.20

556   JCB Annual Reports, 1949-50, p.27; 1951-52, p.25

557   PDTS Vol.1 Section B, p.121

558   PDTS Vol.1 Section B, p.126

559   B Galligan, Utah and Queensland Coal, 1989, p.40

Chapter 9

560   J. Priest, The Thiess Story, 1981, p.139

561   Priest, pp.139-140

562   Prospecting and Developing the Moura Mine, QGMJ February 1968

563   Priest p.168

564   SMH 1 November 1960 p.12

565   QGMJ February 1968, p.51

566   Galligan, p.50

567   JCB Annual Report 1961-62 p.16

568   A. Trengove, Discovery: Stories of Modern Exploration, 1979, pp.28-29

569   Trengove, p.29

570   Trengove, p.31

571   Trengove, pp.31-32

572   Powell Duffryn p.245

573   D. King, QGMJ, May 1969

574   Galligan. P.55

575   Galligan, p.55

576   Trengove, p.44

577   Galligan p.43

578   Galligan p.60

579   Galligan, p.59

580   Galligan, p.59

581   Trengove, pp.41-42

582   Galligan p.62

583   CT 26 July 1968, p.13

584   ibid

585   CT 21 Jan 1969 p.1

586   CT 21 Jan 1969 p.1

587   JCB Annual Report 1959-60, p.50

588   This section is based on SMH articles: 10 June 1963 p.5; 12 June 1963 p.1; 13 June 163 p.2; 27 November 1963 pp.1,5; 28 November 1963 p.5

589   SMH 10 June 1963, p.6

590   SMH, 28 November 1963, p.5

591   SMH 27 November 1963, p1,5

592   JCB Annual Report 1959-60, p.60. The JCB impact was expressed as 850 man years of work.

593   SMH 16June1960 p10

594   SMH 14 July 1959, p.8

595   SMH 18May1960 p.4

596   SMH 28 November 1963, p.5

597   SMH, 17 December 1969, p.20

598   Royal Commission Appointed to Inquire into Certain Matters Appertaining to the State Coal Mine, Collinsville, p.263

599   RC Collinsville. P.260

600   RC Collinsville, pp.263-264

601   QGMJ 12 march 1961, p.132

602   Jay, p.177

603   CT, 16 December 1963, p.15

604   SMH 12 March 1962, p.4

605   SMH 28 June 1961, p.4; 3 November 1961, p.1

606   Jay, p.181

607   SMH 11 September 1962, p.8

608   Sun Herald, 19 June 1960, p.85

609   SMH 8 November 1965, p.1

610   SMH 25 March 1969, p.5

611   Article by Gavin Souter, Sydney Morning Herald, 19 March 1971, p.7

612   SMH 19 March 1971, p.7

613   SMH 6 November 1970, p.5

614   CT 6 October 1971, p.8; SMH 24 June 1972, p.1

615   SMH 27 October 1972, p.2

616   JCB Annual Report 1959-60, p.56

617   D Lee, The Second Rush: Mining and the Transformation of Australia, p.49

618   Lees, p.50

619   CT 28 October 1966, p.3

620   http://adb.anu.edu.au/biography/williams-idris-12033 accessed 20November2017

621   Australian Women's Weekly, 4 December 1968,m p.22

622   F Brady, Electricity Supply in NSW, The First Century and Beyond, p.31

623   Brady, p.31

624   CT 10 October 1964, p.27

625   SMH 2 October 1964, p.4

626   CT 2 December 1965, p.24; SMH 2 December 1965, p.11

627   CT 22 July 1966, p.14

628   Ross, p.505

629 Common Cause 16 December 1967, p.3

630 ibid

631 SMH 30 April 1979, p.3

632 JCB Annual Report 1972-73, p.236

633 NSW Department of Mines Annual Report 1965, p.65

634 NSW Department of Mines Annual Report, p.97

635 NSW Department of Mines Annual Report 1970, p.72

636 SMH 29 May 1957, p.6

637 NSW Department of Mines Annual Report 1951, p.55

638 NSW Department of Mines Annual Report 1958, p.72

639 QGMJ December 1954, p.961

640 NSW Department of Mines Annual Report 1965, p.71

641 NSW Department of Mines Annual Report 1966, p.74

642 NSW Department of Mines Annual Report 1965, p.65

643 NSW Department of Mines Annual Report 1966, p.68

644 QCB Annual Report 1969-70, p.14

645 QCB Annual Report 1961-62, p.6

646 A Murray, No Easy Field, p.264

647 Murray, p.265

648 SMH 28 November 1968, p.1

649 SMH 29 November 1968, p.7

650 JCB 1968-69, p.18

651 SMH 19 March 1969, p.22; SMH 13 May 1970, p.23

652 SMH 20 March 1969, p.20

653 SMH 17 December 1969, p.20

654 QCB data; average employment for years ended June.

Glossary of terms

Anthracite

A hard variety of coal, having a very high carbon content (over 90%) and low moisture content; mainly used for residential and commercial heating. In geological terms, the oldest form of coal.

Bank

An old term used for the area at the top of the pit or mine shaft.

Bank to bank

A term used in labour agreements in the 1800s eg "8 hours bank to bank" meant the length of the working day measured from the time the worker entered the shaft to the time he arrived back at that point.

Bituminous coal

The category or rank of coal that includes most thermal coal and all metallurgical coal. There are sub-bituminous coals which are or were used for power generation (eg from Leigh Creek in South Australia, Callide in Queensland and Collie in Western Australia).

Blast furnace

A furnace in a steel mill in which coke reacts with iron ore to reduce it to iron.

Bord and pillar

The dominant type of underground coal mining in Australia until the 1980s and 1990s when longwall mining became more common. A bord is a "heading" or tunnel or road. The coal is removed as the bords are developed, leaving the pillars of coal to support the roof of the area being mined. Pillars can also be extracted under certain conditions.

Brown coal.

Also called lignite. Contains a lower percentage of carbon than bituminous coals and has a high moisture content. Used in Victoria for power generation. Because of its high moisture content it is unsuitable to transport long distances and so is essentially a domestic fuel in Australia.

Cavil

A ballot which was conducted among contract mine workers before World War Two to allocate work locations in a mine (so as to ensure that all miners worked in areas which were less productive).

Crib room

An underground area where miners eat their meals.

CIF

Cost, insurance and freight. A term used in marketing and transport. When coal is purchased on a CIF basis, the supplier is contracted to arrange and pay for it to be shipped to the customer, with the price including all the costs to the import port, not including any import duties.

Coal

Vegetable matter which has been changed through a process of heat, chemical action and pressure over long periods of time. See also anthracite, bituminous coal, thermal coal, metallurgical (coking) coal, PCI coal and brown coal (lignite).

Coalification

The process whereby heat and pressure turn decomposing plant material to coal over millions of years. Peat is the first stage in this process, but is not regarded as coal. See also Rank.

Coal dust

Fine powdery dust produced by mining which can be explosive under certain conditions. Also dangerous to humans, causing lung disease.

Coal face

The working location where coal is being cut from the seam.

Coalfield

An area or district containing coal. In NSW the major coalfields are the Newcastle, Hunter, Gunnedah, Western and Southern coalfields. In Queensland the major coalfields are commonly referred to as basins, with the North Bowen Basin, South West Bowen Basin, South East Bowen Basin, Surat Basin, Callide Basin and Galilee Basin, plus the Clarence Moreton Basin as the major areas.

Coal handling and preparation plant.

A plant which washes coal to remove impurities and carries out other processes such as screening to produce coal of a certain size. Often simply called a washery.

Coal seam gas

Also called coal seam methane and coal bed methane. The gas which is predominantly methane is contained in seams of coal and is released during mining or by the "fracking" process now common in the USA and in Queensland.

Coal terminal (also called export coal terminal

The facility at a port where coal is delivered from the mine, stockpiled and finally loaded onto the ship.

Coking coal

See Metallurgical coal

Collier

A term now rarely used. Can refer to someone who mined coal or worked in a coal mine or to a ship used to transport coal.

Colliery

A coal mine, including its infrastructure (roads, plant, machinery, equipment and coal washery).

Continuous miner

An electrically powered machine mounted on tracks which has a cutting arm which rips the coal from the face. The coal is gathered by mechanical arms and loaded onto a conveyor. The machine is also used to drive roadways or headings.

Contract miner

A miner who was paid on the basis of the weight of coal he produced. Wheelers who used pit ponies to transport coal skips were also contract mine workers. Other mine workers were day wage workers, hired and paid on a day-by-day basis. The contract mining system existed up until the 1950s.

Custom and practice

The range of informal rules, procedures and standards applying in the coal industry or at different minesites at different times and which at one time were recognised in industry awards and by the Coal Industry Tribunal.

Darg

A limit on the amount of coal each miner could produce in a day; dargs were introduced by the miners' union in the 1800s.

Davy lamp

A safety lamp developed by Sir Humphrey Davy in 1815 which indicated the presence of gas in a mine.

Decline

A sloping tunnel big enough for vehicles, loaders and trucks to drive up and down to get to the coal or ore and take it out.

Demurrage

The cost typically paid by the exporter as a result of port delays. Typical contracts allow for a certain time for a ship to enter a port, load the coal and depart and for payment of a certain amount of demurrage per day if that time is exceeded.

Deputy

A coal industry work classification in NSW. Deputies are underground supervisors.

Dragline

A large machine operated by one person which removes overburden in open cut mines. Draglines have a long arm or boom from which hangs a bucket, with the largest buckets capable of holding over 100 cubic metres of rock and earth.

Drift

An inclined haulage road to the surface. Also a heading used for exploration or ventilation.

Face

Also coal face. The working area where coal is being extracted.

Fire-damp

An old term for gases found in coal mines, methane being the most common, and which become explosive under certain conditions.

Free on board (FOB)

Most Australian coal is sold on an FOB basis where the customer pays for the costs of shipping, insurance, unloading and other costs. The supplier is responsible for costs up to and including the loading of the coal onto the ship at the export port.

Gas coal

Coal was used in Australia from the 1830s to the 1970s to produce town gas ie gas used in urban areas for street lighting, lighting in homes, cooking etc. The Greta seams between Maitland and Cessnock contained some of the highest quality coals suitable for gas making. These coals were also well suited for use in railway locomotives in the era before diesel took over as the preferred locomotive fuel.

Goaf

That area in a mine from which coal has been extracted. The goaf will generally have been filled up by the collapse of the roof of the mine or can be expected to filled once the roof collapses.

Grunching

A term used for the practice in some mines where coal was blasted out of the face by explosives.

Hard coal

A term used for example by the International Energy Agency to include thermal coals, metallurgical coals and anthracite. Also a term for black coal (as distinct from brown coal).

Kg
Kilogram

Lodge

The unit or local branch of the Miners' Federation (now CFMMEU) at individual mines.

Longwall mining

The predominant form of underground coal mining in Australia which is also very safe and highly efficient. Parallel headings or tunnels are driven up to several hundred metres apart and then connected to form the longwall face. A shearer travels along the coal face cutting the coal. The roof is supported by a series of hydraulic supports. As the coal is cut, the shearer and the supports can all move. As the system "retreats" the roof behind the hydraulic supports caves in.

Master

A term used in the 1800s referring to the owner of a coal mine.

Metallurgical coal

The type of coal used in the manufacture of steel and other metals; includes hard coking coal, soft coking coal and PCI coal. See also Steelmaking

Mine subsidence

The sinking or drop in the level of the land surface due to underground mining.

Mt

Million tonnes

Mtpa

Million tonnes per annum

MWh

One megawatt hour equals 1,000 kilowatt hours . A company consuming one megawatt hour of electricity is using 1,000 kilowatts continuously for one hour.

Oil shale

A sedimentary rock which may contain hydrocarbons which can be extracted to produce petroleum products. Oil shale was mined near Lithgow during World War 2 and processed to produce petroleum. It was also mined from the 1850s to produce kerosene for domestic use.

Open cut mining

Mining from an open hole in the ground. Also called open cast mining.

Overburden

The volume of material (rock, dirt etc) removed to uncover the seam of coal (or other mineral) in an open cut mine. Usually expressed in cubic metres. Also known as spoil.

Pit

Can refer to a colliery, or just to the area of an open cut mine being mined, or to the underground workings.

Pit pony

A horse used underground, mainly to haul coal skips.

Pit prop

A support used underground to prevent the roof from collapsing. Usually timber.

Pit-top

The surface area at the top of the mine shaft.

Productivity

The productivity of a mine is a ratio of its output to one or more of its inputs. Common measures of labour productivity are tonnes produced per person per year and tonnes per person per hour. Other measures of productivity can include total factor productivity (ie including labour and capital equipment) and productivity measures relating to major items of equipment such as draglines and longwall systems.

Rank

A method of classifying coal based on the amounts of carbon and volatile matter it contains. It signifies the coalification of the organic material or the degree to which the coal has changed and matured. The lowest rank coal is lignite; next come the sub-bituminous coals; then the bituminous coals (thermal and metallurgical); with the highest rank coal being anthracite.

Reserves

In lay terms, reserves of coal or minerals refer to the quantity of coal or ore which can be mined under certain assumptions. Under the Australian standard for reporting (the JORC code), reserves can be proven or probable. The code defines an 'ore reserve' as the economically mineable part of a measured and/or indicated mineral resource.

Roof bolt

Modern coal mining and metalliferous mining use roof bolts to strengthen the roof of the mine and minimise the likelihood of the roof collapsing. Roof bolts are drilled into the roof and contain chemicals which solidify and anchor the bolt. Roof bolts have replaced props in modern mines throughout the world.

Raw coal

The coal produced from a mine before being washed.

Run of mine coal

Coal as it comes from the mine prior to any screening or other treatment.

Safety lamp

Generally refers to lamps used in coal mines which replaced lamps with a naked flame ie a flame exposed to the air. The Davy lamp was invented by Sir Humphry Davy in the UK in 1815 and it indicted the presence of methane in a mine. The other major safety lamp was the Geordie lamp also invented in 1815 by George Stephenson. Both lamps produced a poor light and were unpopular with miners.

Saleable coal

The coal which is in a form suitable for selling and transporting to the user. Most Australian coal is washed prior to sale (and so changes from raw to saleable coal), but some coal is sold without being washed.

Seam

The layer of coal or mineral. Coal seams can range in thickness from a few centimetres to 30 or more metres.

Shaft

A vertical hole or tunnel which gives access to the mine and can be used for transporting workers and equipment and for ventilation and other services.

Shale

A fine-grained, sedimentary rock composed of mud from flakes of clay minerals and fragments of other materials. Can contain methane gas. See also Oil Shale.

Shearer

A rotating cutting device used in underground coal mining.

Skip

A container or wagon used to transport coal from the face to the unloading point. In the early days of mining skips were wheeled by men or boys or horses to the bottom of the shaft or to the mine exit.

Soft coking coal

A lower quality of coking coal used in steelmaking. In Australia most soft coking coal comes from the Hunter Valley or Gunnedah coalfield.

Steaming coal

See Thermal coal.

Steelmaking

Steel is made through one of two major processes – the blast furnace/ basic oxygen furnace process (BF/BOF) or the electric arc furnace (EAF) process.

In the BF/BOF process coking coal is heated in large ovens to produce coke (which is composed almost entirely of carbon). The coke is then used in the making of iron, with a mixture of coke, iron ore and limestone poured into a blast furnace. The coke burns to produce carbon monoxide which in turn reduces the ore to liquid iron (or hot metal) which has absorbed the carbon from the carbon monoxide. The liquid iron is then passed through a basic oxygen furnace to produce crude steel. On average, the BF/ BOF process uses 1,400 kg of iron ore, 770 kg of coal, 150 kg of limestone, and 120 kg of recycled steel to produce a tonne of crude steel. (World Steel Association).

The EAF process uses primarily recycled steels and/or direct reduced iron and electricity. On average, the recycled steel-EAF route uses 880 kg of recycled steel, 150 kg of coal and 43 kg of limestone to produce a tonne of crude steel. (World Steel Association)

Strip (or stripping) ratio

The relationship between the volume of overburden which has to be removed for each cubic metre of coal. Can also be expressed in other ways eg the ratio of the thickness of overburden to the thickness of the coal seam. A mine with a low strip ratio will typically be able to produce coal more cheaply than a similar mine with a higher ratio.

Stowage

The practice of filling in the area left unsupported once coal pillars have been removed in an underground mine, using rock or other material.

Tailings

The waste material which is left after coal or minerals are washed or otherwise treated prior to sale.

Thermal coal

Also known as steaming coal and energy coal. Coal which is suitable for use in power stations or other plants to generate steam; the steam then drives turbines which generate electricity. Thermal coal is also used in kilns in the manufacture of cement.

Washery

A plant to wash impurities from coal before it is shipped to users. See also Coal handling and preparation plant.

Wheeler

An underground worker who loaded and moved skips.

Bibliography

Bach, J, *A Maritime History of Australia*, Thomas Nelson, West Melbourne, Vic, 1976.

Baddeley, J.M., *Coal*, Acting Government Printer, Sydney, 1943.

Broken Hill Proprietary Company, *Mechanisation in the steel industry's collieries*, Broken Hill Proprietary Company and Australian Iron & Steel, Melbourne, 1949.

Broken Hill Proprietary Company, *Steel industry's colliery development*, Broken Hill Proprietary Company, 1948.

Broken Hill Proprietary Company, *Collieries brochure 1953*, Broken Hill Proprietary Company, 1953.

Blainey, Geoffrey, *The steel master: a life of Essington Lewis*, Melbourne University Press, Carlton Vic., 1995.

Bowden, Bradley (ed), *Work and Strife in Paradise: the History of Labour Relations in Queensland 1959-2010*, Federation Press, Annandale, NSW, 2009.

Bowen, Bruce and Gooday, Peter, *The Economics of Export Controls, ABARE Research Report 93.8,* Australian Bureau of Agricultural and Resource Economics, Canberra, 1993.

Brady, Frank, *Electricity Supply in NSW – The First Century and Beyond* , Institution of Engineers Australia, 1994.

Branagan, David, Geology and Coal Mining in the Hunter Valley, Newcastle Public Library in assoc with Newcastle and Hunter District Historical Society, 1972.

Branagan, David (ed),*Coal in Australia: the Third Edgeworth David Symposium, September1990.*

Broomham, R., *First Light: 150 years of Gas,* Hale & Ironmonger, Sydney, c1987.

Burley, K., *The Overseas Trade in NSW Coal and the British Shipping Industry 1860-1914*, The Economic Record, Vol.XXXVI.

Butlin, S.J., *War Economy 1939-1942*, Australian War Memorial, Canberra, 1961.

Butlin, S.J., and Schedvin, C.B., *War Economy 1942-1945*, Australian War Memorial, Canberra, 1977.

Cameron, Keith, *The Problem of Coal*, Institute of Public Affairs IPA (Victoria), Review, May June 1951.

Campbell, K. L., *Report of the Royal Commission to Inquire into the Coal Industry and Coal Trade in the State of New South Wales*, First Report, Government Printer, Sydney, 1919.

Campbell, K. L., *Report of the Royal Commission to Inquire into the Coal Industry and Coal Trade in the State of New South Wales*, Second Report, Government Printer, Sydney, 1920

Christison, Ray, *A light in the Vale: Development of the Lithgow District Miners' Mutual Protective Association*, City of Greater Lithgow Mining Museum, Lithgow c2011.

Colebatch, Hal G.P., *Australia's Secret War*, Quadrant Books, Balmain NSW, 2013.

Collins, D., *An account of the English colony in New South Wales*, Reed in assoc with Royal Historical Society, Sydney 1975.

Comerford, Jim, *Coal and colonials: the founding of the Australian coal mining industry*, United Mineworkers Federation of Australia, Sydney, 1997.

Cousins, Arthur, *The Garden of New South Wales: a history of the Illawarra & Shoalhaven Districts, 1770-1900*, Illawarra Historical Society, Wollongong, 1994.

Cremin, Aedeen, *the Growth of an Industrial Valley: Lithgow, New South Wales*, Australian Historical Archaeology, 7, 1989.

Culter, Suzanne, *Managing decline: Japan's coal industry restructuring and community response*, University of Hawaii Press, Honolulu, c1999.

Daly, Fred, *From Curtin to Kerr*, Sun Books, South Melbourne, 1912.

Danvers Power, F., *Coalfields and Collieries of Australia*, C. Parker, Melbourne, 1912.

Davidson, C.G.W., *Commonwealth Board of Inquiry appointed to inquire into and report on the coal industry*, Parliament of Australia, March 1946.

Davidson, C.G.W., *Report of the Royal Commission appointed to inquire and report on the coal industry*, Parliament of New South Wales, 1930.

Davidson, C.G.W., *Report of the Royal Commission of Inquiry into safety and Health of Workers in Coal Mines*, Parliament of NSW, 1939.

Davidson, C.G.W., *Coal in Australia*, Fifth Empire Mining & Metallurgical Conference, AusIMM, Melbourne, 1953.

Day, David, *Chifley*, Harper Collins, Sydney, 2001.

Deery, Phillip, *Labor in Conflict: the 1949 Coal Strike*, Hale and Ironmonger, Sydney, 1978.

Deery, Phillip, *Chifley, the Army and the 1949 Coal Strike*, Labour History No.68, May 1995.

Dingsdag, Donald, *the restructuring of the NSW coal industry, 1903-1982*, PHD Thesis, University of Wollongong, 1988.

Dingsdag, Donald, *The Bulli Mining Disaster: Lessons from the past*, St Louis Press, Sydney, 1993.

Dunn, Col, *A History of Electricity in Queensland*, Bundaberg, Queensland, 1985.

Dunne, E.F., *Brief History of the Coal Mining Industry in Queensland*, Journal of the Royal historical Society of Queensland, Vol.4 Issue 3, Brisbane, 1950.

Elford, H. and McKeown, N., *Coal Mining in Australia*, Tait Publishing Company, Melbourne, 1947.

Ellis, M.H., *A saga of coal: the Newcastle Wallsend Coal Company's centenary volume*, Angus & Robertson, Sydney, 1969.

Fisher, Chris, *Coal and State*, Methuan, Sydney, 1987.

Freudenberg, Graham, *Cause for Power: The Official History of the New South Wales Branch of the Australian Labor Party*, Pluto Press, Sydney, 1991.

Galligan, Brian, *Utah and Queensland coal: a study in the micro political economy of modern capitalism and the state*, University of Queensland Press, Brisbane, 1989.

Gollan, Robin, *The Coalminers of New South Wales, A History of the Union 1860-1960*, Melbourne University Press, Melbourne, 1963.

Gollan, Robin, *Revolutionaries and Reformists: Communism and the Australian Labour Movement 1920-1955*, ANU Press, Canberra, 1975.

Gollan, Robin, *Radical and Working Class Politics*, Melbourne University Press, Kingsgrove NSW, 1970.

Gregson, Jesse, *The Australian Agricultural Company, 1824-1875*, Angus & Robertson, Sydney, 1907.

Gunn, John, A*long Parallel Lines: A History of New South Wales Railways*, *Melbourne University Press, Carlton Vic., 1989.*

Hargraves, A.J. and Martin, C.H. (eds) *Australasian coal mining practice*, AusIMM, Parkville Vic., 1993.

Hasluck, Paul, *the Government and the People 1942-1945*, Australian War Memorial, Canberra, 1970.

Hasluck, Paul, *the Government and the People 1939-1941*, Australian War Memorial, Canberra, 1952.

Hawthorne, W.L., *Coal exploration by the Queensland Department of Mines during the period 1950 to 1984*, Queensland Geological Record, 2011/07.

Howard, John, *The Menzies Era*, Harper Collins, Sydney, 2014.

Hughes, Helen, *The Australian Iron and Steel Industry 1848-1962*, Melbourne University Press, Melbourne, 1964.

Jack, R.P., *A Review of the Coal Position*, AusIMM, Sydney, 1948.

Jay, Christopher, *The Coal Masters: The History of Coal and Allied 1844-1994*, Focus Publishing, Sydney, 1994.

Joint Coal Board, *Review of Activities*, JCB, Sydney, 1962.

Joint Coal Board, *The competition between coal and oil*, JCB, Sydney, 1964.

Joint Coal Board, *Joint report on a survey of labour requirements in the New South Wales coal mining industry*, JCB, Sydney, 1951.

Kemp, R.J., *Report on Large Scale Development of the Blair Athol Coalfield*, Department of the Coordinator General of Public Works, Queensland, June 1947.

Killin, Kerry, *Drovers, Diggers and Draglines: A History of Blair Athol and Clermont*, Pacific Coal, Brisbane, 1984.

Kinninmonth, R.J. and Baafi, E.Y. (eds), *Australasian Coal Mining Practice*, AusIMM, Carlton Vic., 2009.

Kume, Ikuo, *Disparaged Success: Labour Politics in Postwar Japan*, Cornell University Press, Ithaca NY, 1998.

Lang, J.T., *The Great Bust: the depression of the thirties*, Angus and Robertson, Sydney, 1962.

Lee, David, *The Second Rush: Mining and the transformation of Australia*, Connor Court Publishing, Redlands Bay Qld, 2016.

Merrett, D. (ed), *Business Institutions and Behavior in Australia*, Frank Crass, London, 2000.

Metcalfe, Andrew, *For freedom and dignity: historical agency and class structures in the coalfields of NSW*, Allen & Unwin, Sydney, 1988.

Martin, C.H. et. al., *History of Coal Mining in Australia*, AusIMM Monograph No.21, 1993.

Mauldon, R.F.E., *The economics of Australian coal*, Melbourne University Press in assoc with Macmillan, Melbourne, 1929.

Monger, N.R., *The Newcastle coalfield- 100 years of change*, AusIMM Centenary Proceedings, 1993..

Mumford, K., *Structure and Disputation in New South Wales coal industry, 1952-1987*, Australian Economic History Review, Vol.34 No.1, 1994.

Murray, Alan, *No Easy Field: Ipswich coal mining 1920 to 2000*, University of Queensland Press, St. Lucia Qld, 2010.

Murphy, D. (ed), *The Premiers of Queensland*, University of Queensland Press, St. Lucia Qld, 1990.

Newcastle & Hunter District Historical Society, *Minmi, The Place of the Giant Lily*, Newcastle, 1972.

NSW Parliament, *Progress Report from the Joint Committee of the legislative Assembly upon the Coal Industry*, Government Printer, Sydney, 1964.

National Health Survey Committee, *Report on the Health of coal-miners*, Government Printer, Canberra, 1946.

Norman, A.J., *Black Diamonds: the story of coal in Queensland*, Queensland Coal Board, Brisbane, 1963.

Orr, W., *Mechanisation: Threatened Catastrophe for coalfields*, Miners' Federation of Australia Sydney, 1935.

Orr, W. and Nelson, C., *Coal: the struggle of the mineworkers*, Central Council Miners' Federation of Australia, Sydney 1935.

Patience, Alan (ed), *The Bjelke Petersen Premiership 1968-1983: issues in public policy*, Longman Cheshire, Melbourne, 1985

Piggin, Stuart, and Lee, Henry, *The Mt Kembla Disaster*, Sydney University Press in assoc with Oxford University Press, Melbourne, 1992.

Powell Duffryn Technical Services, *First Report on the Coal Industry of Queensland*, Government Printer, Brisbane, 1949.

Priest, Joan, *The Thiess Story*, Booralong Publications, Ascot Qld, 1981.

Queensland Department of Coordinator General of Public Works, *Report on Large Scale Development of Blair Athol Coalfield*, Government Printer, Brisbane, 1947.

Reeves, Andrew, *Up from the Underground: Coalminers and Community in Wonthaggi 1909 to 1968*, Monash University Publishing, Clayton Vic., 2011.

Reynolds, John, *Men and Mines: A History of Australian mining 1788-1971*, Sun Books, Melbourne, 1974.

Richards, J., *Report of the Inquiry into Recent Mechanisation and Other Technological Changes in Industry*, Government Printer, Sydney, 1963.

Ross, Edgar, *the Coal Front: An Account of the 1949 Coal Strike and the Issues it Raised*, Miners' Federation of Australia, Sydney, c1950.

Ross, Edgar, *A History of the Miners' Federation of Australia*, Australasian Coal and Shale Employees' Federation, 1970.

Ross, Edgar, *Of storm and struggle: pages from Labour history*, Alternative Publishing Cooperative, Sydney, 1982.

Ross, Lloyd, *John Curtin: a biography*, MacMillan, South Melbourne, 1977.

Shaw, A.G.L. and Bruns, G.R., *The Australian Coal Industry*, Melbourne University Press, Melbourne, 1947.

Sheridan, Tom, *Division of Labour, Industrial Relations in the Chifley Years*, Oxford University Press, Melbourne, 1987.

Spires, Robert, *History of Kemira Colliery 1857-1984*, Wollongong City Council Library.

Sumiya, Mikio, *A History of Japanese Trade and Industry Policy*, Oxford University Press, New York, 2000.

Taylor, Ross and Murray, Alan, *In the Black: The Auscoal Super Story*, Auscoal Super, Newcastle, 2011.

Thomas, Pete and Gorman, Paddy, *The coal mines that workers ran: from Nymboida to United a remarkable chapter in Australia's industrial history*, CFMEU Mining & Energy Division, Sydney, 2000.

Trengove, Alan, *Discovery: Stories of Modern Mineral Exploration*, Dominion Press, North Blackburn Vic., 1979.

Trengove, Alan, *What's good for Australia..! the story of BHP*, Cassell Australia, Stanmore NSW, 1975.

Tsokhas, Kosmas, *Beyond dependence: companies, labour processes and Australian mining*, Oxford University Press, Melbourne, 1986.

Turner, J.W., *Coal Mining in Newcastle, 1801-1900*, Council of the City of Newcastle, Newcastle NSW, 1982.

Turner, J.W. (ed), *Newcastle as a Convict Settlement: Tte Evidence before J.T. Bigge in 1919-1821*, Council of the City of Newcastle, Newcastle NSW, 1973.

Vines, Jack, *Coal Mining Heritage Study in Victoria*, Heritage Council of Victoria, Melbourne, 2008.

Walker, Alan, *Coaltown: a social survey of Cessnock*, Melbourne University Press, Carlton Vic., c1945.

Watt, Alan, *The Evolution of Australian Foreign Policy 1938-1965*, Cambridge University Press, London, 1967.

Whitmore, R.L., *Coal in Queensland, the first 50 years: a history of early coal mining in Queensland*, University of Queensland Press, St. Lucia Qld, 1981.

Whitmore, R.L., *Coal in Queensland the late nineteenth century 1875 to 1900*, University of Queensland Press, St. Lucia Qld, 1985.

Whitmore, R.L., *Coal in Queensland: from Federation to the twenties 1900 to 1925*, University of Queensland Press, St. Lucia Qld, 1991.

Wilkenfeld G. and Spearitt P., Electrifying Sydney – 100 years of Energy Australia, Energy Australia, Sydney, 2004.

Windross J. and Ralston J.P., *Historical Records of Newcastle, 1797 – 1897*, Federal Printing and Bookbinding Works, Newcastle NSW, 1978.

Wright, Michael, *Muted Sirens: Demarcation and Union Coverage in the Australian Coal Mining Industry*, ACIRRT Working Paper 27, May 1993.

Index